T0193037

BITTER HEAT

Deliberate Global Warming via 'Trick' Cheney, Texas Oil, Spies, Organized Crime and Big Real Estate

Roger Phelps

Bitter Heat: Deliberate Global Warming Via 'Trick' Cheney,Texas Oil, Spies, Organized Crime and Big Real Estate

Published by:
Trine Day LLC
PO Box 577
Walterville, OR 97489
1-800-556-2012
www.TrineDay.com
trineday@icloud.com

Library of Congress Control Number: 2022945121

Phelps, Roger, —1st ed. Bitter Heat
p. cm.

Epub (ISBN-13) 978-1-63424-412-1
TradePaper (ISBN-13) 978-1-63424-411-4
1. Climatic changes -- Africa, Sub-Saharan. 2. Global warming -- Political aspects. 3. Climatic changes -- Economic aspects. 4. World politics. 5. Social Justice 6. Petroleum -- Environmental aspects.. I. Roger Phelps II. Title

FIRST EDITION
10 9 8 7 6 5 4 3 2 1

Distribution to the Trade by:
Independent Publishers Group (IPG)
814 North Franklin Street
Chicago, Illinois 60610
312.337.0747
www.ipgbook.com

Publisher's Foreword

Mother, mother
There's too many of you crying
Brother, brother, brother
There's far too many of you dying
You know we've got to find a way
To bring some lovin' here today, yeah
–Marvin Gaye, "What's Going On"

Well, I'm about to get sick
From watchin' my TV
Been checkin' out the news until my eyeballs fail to see
I mean to say that every day is just another rotten mess
And when it's gonna change, my friends, is anybody's guess
So I'm watchin' and I'm waitin'
Hopin' for the best
Even think I'll go to prayin'
Every time I hear 'em sayin'
That there's no way to delay that trouble comin' every day
No way to delay that trouble comin' every day
–Frank Zappa, "Trouble Every Day"

Sometimes you wake up in the morning wondering what are you going to do. Other times you know before you go to bed. TrineDay has been a labor of love and duty. Love because I love books, and have learned so much from them, duty because, as I said so many times, my daddy told me some stuff I didn't understand many years ago, and it is my duty to report that, and what it portends, to my fellow passengers of this spinning blue ball.

Roger Phelps is doing his duty by informing us of very interesting facts and actions that are happening but with very little attention paid. I heard from various readers of Roger's first book, *Not Exactly the CIA: A Revised History of Modern American Disaster* about how they were blown away by the information they found and now understood our history better. One internationally-know investigative reporter/author thanked me profusely then saying, "Wow, your book came at the right time for my research and shows me I still have a lot to learn."

Roger's new book, *Bitter Heat: Deliberate Global Warming Via 'Trick' Cheney, Texas Oil, Spies, Organized Crime and Big Real Estate* again shows his ability to get ahead of, and cover vital news stories that no one else is talking about. TrineDay is proud and humble to publish this slim volume that has the ability to upset apple carts – if folks will read and if it can garner some attention.

Sadly, TrineDay can't even get a Wikipedia page, several folks have tried to set one up for us, but they have been unable to get past the censors, who claim that our books are insignificant and not relevant. TrineDay has been in operation for over twenty years and have over 160 books in print!

Russ Baker of WhoWhatWhy.org, and author of *Family of Secrets: The Bush Dynasty, America's Invisible Government, and the Hidden History of the Last Fifty Years* has been writing about his experience with the "real" cancel culture: how his book, from a major publisher was promised rave mainstream coverage that didn't happen and he soon found himself on the radio talking to Alex Jones.

In a recent piece, Russ quoted from a response he got from David Talbot, author of *The Devil's Chessboard: Allen Dulles, the CIA, and the Rise of America's Secret Government,* and co founder of Salon.com, "During my press tour, a book editor at the *Washington Post* told my publicist, 'We won't touch this one with a ten-foot pole.' ... No major TV program or publication featured me or my book, despite my long track record in the media and publishing. I was clearly blacklisted."

TrineDay is "blacklisted," our books get no coverage in the press. So it is up to us to tell our friends and neighbors about this very import book. Bitter Heat tells a terrifying story one we must hear: Climate change is not a hoax!

Onwards to the Utmost of Futures!
Peace,
R.A. "Kris" Millegan
Publisher
TrineDay
August 22, 2022

Quis custodiet ipsos custodes?

Contents

Preface

I wrote this book because I hoped to show "how did we get here." How did we get to the crisis of a struggle for life by our fair planet Gaia, Earth, against her adversary, human-caused global warming?

The project began when I noticed a Texas oil company had made a massive natural-gas strike in coastal Mozambique. This seemed to match what must have been envisioned by the U.S.'s 2001 Energy Task Force, which purported to chart America's petroleum future. Its membership was secretly organized by Vice President Dick Cheney.

Around the Mozambique gas prospect emerged figures in the intelligence field (spies) and in organized crime. I noticed to what a great degree their interests ran together in Mozambique – these oilmen, these spies, and these gangsters. All had the motive of securing maximal profit – as opposed to reasonable profit – out of coastal Cabo Delgado, where the gas was, and where "maritime security" was a hot issue, and where the heroin routes were.

It then became clear that all maximal-profit-motivated activities, in Mozambique, and around the planet, accelerate global warming. Since these actions are deliberate, the warming they cause is in a sense deliberate as well.

It is now evident that global warming, incorporating foreign exploitation of African resources, will likely shoot us past the 1.5-degrees-Celsius cap on warming settled on in the 2015 Paris Climate Agreement. This book is an attempt to answer the question, how did that happen? The "it happened naturally" position is no longer tenable. Something human, some set of human interactions, clearly is responsible.

So, which ones? Human interactions frequently display exercises of power. This is true on all levels – from personal to national to transnational. As noted by philosopher Michel Foucault, human exercises of power tend to show themselves as disciplinary.

This book attempts to pick out places where power is articulated in ways that accelerate global warming. The record shows a phenomenon of tropical nations being disciplined into accepting Western consumerism.

So who or what is the power actor doing this disciplining? It showed itself as a composite acting as a unit of Big Oil, spies, organized crime, and Big Real Estate. Together, they have represented a single, powerful player on the world stage.

The philosopher Arthur Schopenhauer posited that in the visible world a seeming heterogeneity or plurality could "represent" an underlying real unity, which he called a single "will," that was subterranean.

For this book, Big Oil-cum-spies-cum-organized crime-cum-Big Real Estate together represent a single will – in the sense of a "will to power" as postulated by Friedrich Nietzsche, Schopenhauer's successor. We will see how this Western will to power has disciplined and subjugated African nations into Westernization, in a phenomenon that deliberately and rapidly accelerates global warming. The results of such warming are coastal lands opened up, by attendant Sea Level Rise, for Western real-estate development. The same phenomenon shows itself in other tropical regions, as well.

Introduction

In a way, this is a tale of the grotesque, the monstrous.

If we humans are "the rational animal," is a lack of reason monstrous?

Not quite. That said, an overarching theme in this book is the difference between a reasonable profit and a maximal profit. Not so much the moral difference as the practical difference – that difference being that *wherever there is an insistence on maximal profit versus reasonable profit, global warming is accelerated*

For humans to qualify as monstrous, what is necessary besides lack of reason?

By definition, a monster is "an extremely wicked person," who as such performs acts "intended to or capable of harming someone or something." That chimes with actions by a system of industries/quasi-industries including organized crime, Big Oil, Big Real Estate, and the spy community. *These characteristic actions, motivated by securing of maximal profit, together obviously either are intended to, or are at any rate capable of, accelerating global warming, and thus harming a planet, Earth. This happens most characteristically in places where some valuable resource is not tightly controlled by government.* Really, it is somewhat similar, on a grand scale, to the smuggling of low-tax New Jersey cigarettes into high-tax New York – governments are not set up to control this particular resource-extraction-for-maximal-profit.

As we shall see, the resource of Mexican oil was not tightly controlled, and so it was exploited in concert by the Los Zetas cartel and Texas Oil. What is it that characterized this exploitation as well as others described in this book?

It was human ruthlessness.

Financier Michael Milken, for example, and Los Zetas cartel head Miguel Trevino Morales, and Merrill Lynch's Stanley O'Neal all "rose to the top" primarily because they were ruthless men. O'Neal's career and tenure as Merrill CEO exemplify the combination of recklessness, short-sightedness, ruthlessness and greed that has become the hallmark of those who have risen to the top of the U.S. corporate establishment over the past quarter century—a period that has seen an unprecedented redistribution of wealth from the working population to a financial aristocracy that wallows in previously unheard of personal wealth.

Another who did this was Donald Trump, responsible arguably for the preventable deaths of some 300,000 Americans by Covid-19 *and responsible certainly for increased methane releases* (after he loosened the regulation of methane emissions).1

And another was Dick Cheney.

In light of the maximal-profit motive held by all these men and those like them, *is it beyond such men, morally, to deliberate on intentionally speeding global warming, by accelerating resource extraction, to create an arena – a vastly warmed earth – that is even more conducive to maximal profit?* Is it beyond them? Why should we think so?

And in the end, as we have seen, any deliberate action capable of causing harm counts the same – practically, legally, and ethically – as deliberate intent to do harm.

Consider the fact that two extractive industries, oil and coal, banded together to pay for pseudo-science and media-manipulative denials of climate change that successfully lulled the public to disbelieve in global warming.

The record suggests more serious collusion occurred as well. In Chapter Six we shall see that the Deepwater Horizon oil disaster of 2010 in the Gulf of Mexico may have been no accident. The record shows *actions by oil industry leaders Halliburton and Transocean made certain that an oil blowout would occur at BP's Deepwater Horizon rig,* and that there was motive for this. Granted, this seems unthinkable at first glance. But do we assume that the blowout at DH must have been an accident? Why, precisely, should we assume this? Because oil barons' ethical record is pristine? Of course their ethical record is not pristine, not by a long shot. To repeat, though, the public record shows, and we shall see, that actions at Deepwater Horizon by Halliburton and Transocean made it fully certain a blowout would occur there – that is, *it was only a matter of time,* and the blowout when it came vastly accelerated global warming.

Finally, planet-wide – especially as the Covid pandemic wanes – the record shows a tremendous, profitable churning of coastal real estate as global-warming-is causing Sea Level Rise, which means that owners of

1 NPR, August 13, 2020.

land near but not on current waterlines soon will own "the new coastline," tremendously more valuable.

I believe that currently by most people, the developments described above are viewed as unconnected. I hope that this book will serve to connect them, in the mind of readers who will then use evidence laid out herein as "placard points" in Climate Marches on Washington.

Lastly, what about China? This book largely emphasizes corporate-capitalist extractive industries from the West in the acceleration of global-warming. I believe this is justified. But of course, China's state capitalism also heavily exploits natural resources in African nations, accelerating global warming. The book's final chapters will address this point of tension.

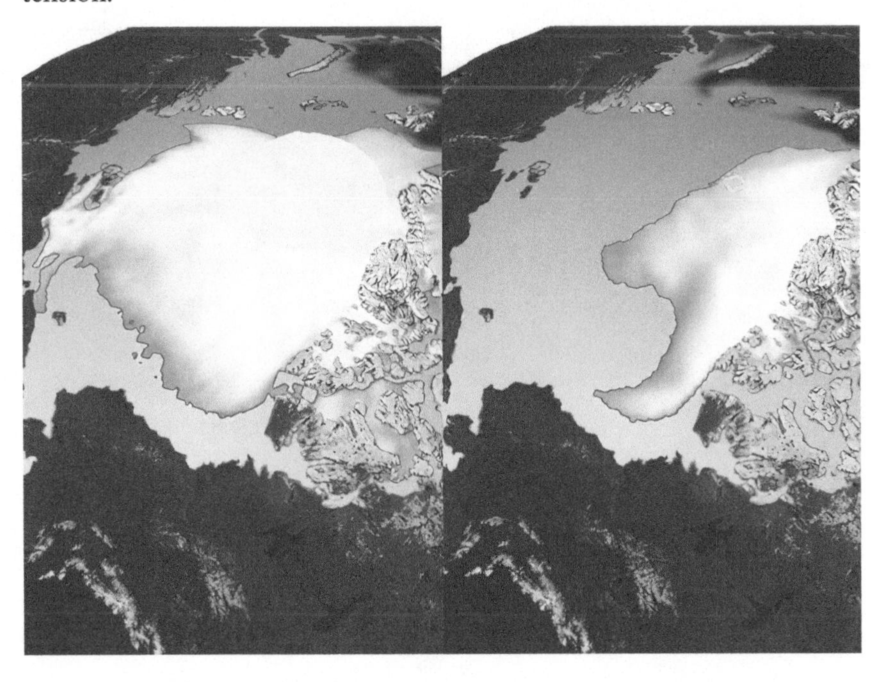

A NASA study published in the *Journal of Climate* shows that the oldest and thickest Arctic sea ice is disappearing at a faster rate than the younger, thinner ice at the edges of the Arctic Ocean's floating ice cap. The images above show the ice cap in 1980, left, and in 2012, right.

Chapter One

MOZAMSCAM: BIG OIL, SPIES FOR SECURITY, AND ORGANIZED CRIME

A U.S.-planned "Gas City "in Mozambique will accelerate global-warming emissions from that country by fully 10 percent.

Vice President Dick Cheney got the press to say his 2001 Energy Task Force, staffed with Texas oilmen from Exxon, Anadarko, and Cheney's Halliburton, arose with the California energy crisis,[1] but this was disingenuous. As the record later showed, the Task Force's energy concern lay abroad, in southern Asian countries and in African countries, including Mozambique. The energy concern was, how best could Big Oil secure maximal profit from these countries' petroleum? The Task Force oilmen were concerned with expanding U.S. fracking, which they got the press to refer to as "to fight reliance on foreign oil."[2]

And of course, California's "energy crisis" had been caused in the first place by Cheney-Bush cronies Ken Lay and Jeff Skilling at Enron.

In this book, "Big Oil" means both large corporations and large drilling operations.

"Security" applies when Big Oil needs to secure a drill permit; which typically requires bribing of foreign officials, by spies, and when Big Oil

1 E.g., *Rochester Business Journal*, January 30, 2001

2 Cf. CBS MarketWatch, May 16, 2002, "Enron Linked to California blackouts"

needs to secure, or guard, a drill operation, which in Africa often requires gang-related armed men.

"Spies" includes those good at securing local cooperation, both from public officials and from private powers – often gangsters – who control turf containing oil deposits. Bribe money from spies works on officials and on gangsters. Spies serving empire is nothing new – historically, it has been the rule. The 19th-century "Scramble for Africa" employed agents from Portugal's Mozambique Company, England's East Africa Company, and the East India companies of Holland, France, and Germany.

This chapter applies this framework to northeastern Africa, where much natural-gas lies offshore, and examines how the concern "maritime security" was easily promotable among African officials, by Western intelligence operatives. These included self-proclaimed CIA asset Erik Prince, founder of Blackwater, and were tasked with securing a maximal profit from Mozambican gas deposits – maximal profit through a *Gas City* built by Big Oil, including Texas outfits Exxon, Anadarko, and Dick Cheney's[3] Halliburton.

Maximum exploitation here means Western companies (including real-estate development companies) are to usurp all benefit from the valuable natural gas, minus a bit for a few complicit Mozambican officials, while locals lose coastal lands and gain nothing except cultural Westernization.[4] As a result, this MozamScam, the record strongly suggests, is responsible for what the Western press calls an "insurgency" in northern Mozambique that, as will be discussed later, in reality is a resources war, pitting an urban ruling elite – aligned with Western oil companies and their intelligence operatives – against a rural population who are not benefiting from development of the prime national resource, petroleum. This rural coastal population instead is being forcibly displaced from villages, without compensation, by a combination of oil exploration and Sea Level Rise brought by an accelerated global warming that itself is a product largely of the oil industry dominated by Western nations. By 2019, arriv-

3 On February 1, 2020, after an absence, Cheney returned as Halliburton chairman. For the previous period and for purposes of this book, note the following: ABC January 6, 2006 A review of regulatory filings showed Cheney is Halliburton's biggest individual stockholder with an estimated $45.5 million stake in the firm. Cheney said that if elected, he'd put his Halliburton holdings in a blind trust. Assuming he did this, there is no reason to believe he left them there after leaving office. And, since there is nothing I could find on the public record about his selling his Halliburton holdings after removing them from a blind trust, it is reasonable to assume he still has them; if so, Cheney likely remains Halliburton's biggest individual stockholder. Even if not, his history with the company – including the millions of dollars in government contracts he steered to Halliburton -- combined with his position as a former U.S. Vice President, allows Cheney fairly to be termed a "controlling" power in the company.

4 BBC, September 8, 2020, "Mozambique's jihadists and the 'curse' of gas and rubies."

ing largely by sea, fighters would capture the port town of Mocimboa da Praia, key to Western natural-gas exploitation in Mozambique.

Nations dominating the oil industry of course include heavily Westernized and oil-rich Arab states, some of which played a crucial role in the MozamScam.

United Arab Emirates

The MozamScam needed the UAE as a kind of support base.

Helping set this up was a move just after 9/11 by George Tenet, the Cheney-Bush CIA director. Tenet "embedded" four CIA agents with the New York City Police Department Counterterrorism Bureau. Then, the United Arab Emirates sought help from this NYPD counterterrorism unit (which employed the four CIA agents). This established an NYPD bond with spies working for the UAE. The UAE in 2008 even paid for this hybrid CIA/NYPD unit to set up an office in Abu Dhabi.[5] Results of this move were never under Tenet's full control.[6] During his embedding at NYPD, CIA agent Larry Sanchez developed "an ongoing relationship" with high-level Emirati officials.[7]

On the basis of this "not-exactly-CIA"[8] operation with UAE spies, quickly entered American Erik Prince, founder of Blackwater and a self-proclaimed CIA "asset." Prince moved to Abu Dhabi in 2009. There, Prince was able to trade on relationships established by Sanchez with powerful UAE sheikhs – in a CIA "dream," as one ForeignPolicy source put it, "to help the UAE create its own CIA." Prince with Sheik Mohamed bin Zayed al Nahyan (MBZ) created "an elite counterterrorism" unit.[9]

In Abu Dhabi, this spy work certainly caught the attention of shipbuilder Iskandar Safa, yacht purveyor to MBZ's relative Hamdan bin Zayed al Nahyan; at this time, Safa's Privinvest, which happens to build mostly "maritime security" boats, employed a spy from South African Defense Department intelligence. More to come on her.

As such, for Safa, Prince, and Cheney in the UAE, interests clearly matched and power bases lay close together.

Given this, and the massive profits foreseen by Texas oilmen in Mozambique, it would have been surprising if MozamScam did not arise.

5 UAE intelligence gifted the New York Police Foundation with millions of dollars to enable "the NYPD[-/CIA] to station detectives throughout the world." Cf. The Gothamist Web site Oct 24, 2019

6 According to a CIA Inspector General's investigation. Cf. *New York Times*, June 26, 2017, and CNN, June 27, 2013. This investigation found "inadequate direction and control" by CIA managers "responsible for the relationship."

7 ForeignPolicy Web site, "Deep Pockets, Deep Cover"; December 21, 201

8 Cf. my *Not-Exactly-the-CIA: A Revised History of Modern American Disasters*, Trine Day, 2019-20

9 *New York Times*, May 14, 2011

In a crafty ploy using dishonest Mozambican officials, Erik Prince wound up owning Mozambique's sole means of maritime security for its offshore gas fields – namely, a fleet of Iskandar Safa-made coastal-patrol boats. This remarkable move was important (coming just at a time when Exxon and Halliburton were poised to begin drilling work in Mozambique's coastal gas fields), because Horn of Africa waters were loudly being called "insecure" – due, it was said to "Somali pirates" – when in reality a massive heroin trade plied the waters, aided by the same dishonest Mozambican officials as were being bribed in the scam. It was this sensitive situation, in which the oilsters needed secretly to get in bed with the gangsters, that needed to be "secured." As we shall see, this took some doing.

Necessary first was concealing what kind of boats Safa had contracted to build. Conveniently, a pliant U.S. press wrote a story referring to "tuna boats" – Mozambique is supposed to have contracted with Safa for tuna boats and then failed to repay a "loan" from Credit Suisse purported to cover the purchase price for tuna boats. That inaccurate account has become known as the "Mozambique hidden-debt scandal"[10] story.

Under that cover surface, the scam involving Prince, Safa and Co. is still playing out.[11] As far as I have been able to determine, the Prince-owned Safa-built boats haven't yet been deployed.[12] But the natural-gas deposits are still there, awaiting developments. If and when that gas is drilled and burned, global warming will accelerate markedly.

Cheney in the background with UAE help

In 2006, as Halliburton was planning its move to Dubai, a company called Dubai Ports World owned Mozambique's Maputo Harbor. That year, DP World also acquired ownership of New York Harbor.[13]

When an anti-Arab Congress pressured DP World to divest the New York port, the Cheney-Bush administration backed DP World, against

10 After the so-called "hidden-debt" affair, the head of logistics at Safar's Privinvest opened a new branch specialized in gas operations in Cabo Delgado Province, Mozambique. AfricaIntelligence Web site November 17, 2020..

11 As we shall see, the record strongly suggests this protracted scheme has now produced in northern Mozambique a civil war, called an "insurgency," a situation into which foreign military forces are poised to enter, with no secret made that their role would be to secure the gas-drilling fields in Cabo Delgado where Western Big Oil has secured permission to drill.

12 As of this writing, it might seem that destabilization in Mozambique has gone too far – a civil war, usually called an insurgency, has stalled petroleum development. But with their wealth, oil corporations can play a waiting game here – as of May 2021, the U.S. already had some troops in Mozambique and the general clamor was for more foreign military intervention under "humanitarian" guise. As it happens, a fledgling "Gas City" in Mozambique's Cabo Delgado is a perfect operating base for military combat and occupation, and as soon as that were to happen, Western gas extraction would proceed, even less regulated than planned.

13 Wikipedia

Congress.[14] The next year, Cheney moved Halliburton to Dubai. A year after that, DP World became official "partner" with Cheney-Bush's Department of Homeland Security, in a deal called C-TPAT ("Customs Trade Partnership against Terrorism"). With this status in hand, DP World in 2010 lucratively got Mozambique to extend DP World's Maputo port ownership role until 2043.[15]

Between Cheney-friendly DP World and Iskandar Safa's Privinvest shipbuilding company, no corporate connection exists. But in 2011, Privinvest provided a super-yacht for Sheikh Hamdan bin Zayed bin Sultan Al Nahyan, son of the UAE founder. The deal came shortly after CIA asset Erik Prince headquartered in the UAE. It came also shortly after Texas's Anadarko secured drilling permission in Mozambique.

Erik Prince

Prince already had a relationship with Mohammed bin Zayed al Nahyan (MBZ) – crown prince of Abu Dhabi – since 2009, when Prince sold the sheikh on creating an elite counterterrorism unit.[16] Through this relationship Prince had credibility for access both to shipbuilder Safa, whose Privinvest company partnered at Abu Dhabi Mar shipyards with MBZ's relative Hamdan bin Zayed al Nahyan, and to officials of the UAE's intelligence agency, National Electronic Security Authority.[17]

Through these relationships and/or in Abu Dhabi social/business circles,[18] Prince is practically certain to have met both fellow Christian and Safa top lieutenant Jean E. Boustani and Boustani's fellow Privinvest employee Basetsana Thokoane. Basetsana also worked at the time for the South African Defence Department intelligence wing.[19]

Privinvest submitted a project for the protection of Mozambique's Exclusive Economic Zone (EEZ) to the office of the Mozambican president on December 31, 2011. Thirteen months of negotiations followed, capped by Boustani's and Thokoane's visit to President Armando Guebuza and other high officials in January 2013.[20]

14 *New York Times*, February 1, 2006
15 AllAfrica Web site, September 20, 2020
16 *The Intercept*, May 3, 2019
17 Since renamed Signals Intelligence Agency
18 By 2011, the U.S. Navy was buying boats from Safa to train Mozambican sailors in maritime security. August 9, 2018, AllAfrica.com Web site
19 Cf. "South Africa: The Post-Apartheid Decade," Centre for Conflict Resolution, 2004.
20 .Hue and Cri Web site, November 20, 2019

In the above photo, the Portuguese-language inset translates as "Basetsana Thokoane will have presented Boustani to Mozambicans."

Together, Boustani and Thokoane lobbied for a Safa boat-building contract with Mozambican officials Armando Guebuza, then president, and Filipe Nyusi, future president. Very likely also present was Antonio Carlos do Rosario, head of SISE, the Mozambican spy agency, plus Armando Guebuza and Filipe Nyusi, former and current Mozambican presidents respectively, with Iskandar Safa, CEO of Privinvest.

What kind of boats? This has been disputed, and thus obfuscated. Reuters, December 13, 2017, reports it accurately, except for referring to the boats as "fishing boats." The *Daily Mail*, London, on April 30, 2016, wrote, "Britain gave £84 million a year to Mozambique to set up fishing fleet... but instead country bought military patrol boats...." Multiple reports have stated, "36 patrol boats and 24 tuna boats." The latest reference I know of, by Bloomberg's News24 Web outlet, says, "$2 billion for a coastal patrol force and tuna fishing fleet."

Ultimately, American Erik Prince bought some or all of these boats. Prince has said he would patrol Mozambique's fishing seas to prevent poaching. These are the same seas that cover the offshore gas deposits coveted by Western oilmen.

Boustani was accused of paying bribes to Mozambican officials and thereby defrauding U.S. investors who had bought some of the loans from Credit Suisse. He never denied paying Mozambique officials, but prosecutors failed to prove that he was guilty of defrauding U.S. investors. Safa was also charged and then acquitted.

Eventually, CIA asset Prince bought the Safa-built patrol boats from a Mozambican state company led by SISE head spy Carlos do Rosario.

To spell it out in a single sentence, the record strongly suggests that in order to ease Texas oil's path in coastal Mozambique, spies from several nations planned to have Prince gain control over coastal-patrol boats tabbed for Mozambique, because – in that the shore of the Arabian Sea-Indian Ocean from Pakistan to Africa is "The Heroin Coast" –[21] any adequate "security" for Texas oilmen's operating offshore there entailed a covert monitoring and a skimming of the maritime heroin trade.

Guebuza Cronies and Heroin

When the Texas oilmen's gas push began, Mozambique's president Armando Guebuza (nickname "GueBusiness") had as cronies several wealthy tycoons – in hotels and import-export – who dealt drugs on the side.[22] This was common knowledge at the U.S. State Department, as the following leaked diplomatic cable shows:

> "In the licit economy (Guebuza) and Mohamed Bashir Suleiman ... demand a cut of all significant business transactions; in the illicit economy, MBS dominates money laundering and drug transshipment, providing kickbacks to (Guebuza's ruling party)," a prominent source told the U.S. embassy. "MBS is the most important narco-trafficker in the country (and has) a money laundering network centred on the family business Grupo MBS."[23]

At Maputo Port – operated, of course, by Dubai Ports World – Suleiman was said to have his own clearing agent, who allowed Suleiman's shipments to bypass the port's mandatory scanning; in four years as Guebuza's Customs Director, Domingos Tivane had amassed a million-dollar fortune by "facilitating drug shipments" to Maputo Harbor, according to Wiki-leaked U.S. Diplomatic Cable 09MAPUTO1291. Beginning in 2010, huge sums of money flowed from Mozambique to Dubai, which with no national regulation on banking or real-estate *is a major money-laundering location.* In March 2013, Dubai Ports World's Chairman, His Eminence Sultan Ahmed Bin Sulayem, gave Guebuza a royal welcome at DP World's flagship Jebel Ali Port.[24]

21 "The Heroin Coast, a political economy along the eastern African seaboard," GlobalInitiative Web site, July 2, 2018, by Simone Haysom, London School of Economics.
22 Ibid. to 5
23 Afrol News, December 10, 2010, "Mozambique drug barons 'protected by President,'" citing wires from the U.S. embassy in Maputo from 2009 and early 2010, published by Wikileaks
24 DredgingToday Web site, Mar 26, 2013..

In June 2001 (while Dick Cheney's secret Energy Task Force was meeting), drug dealing was the Number One economic driver in Mozambique. Mozambican Stock Exchange Chairman Jussub Nurmamade said drug dealing was responsible for stock exchange's rapid growth, by $100 million during just the first half of 2001.

Nurmamade said,

> (This is) unique. How can an economy as small as Mozambique's grow by $100 million in such a short time? What makes Mozambique 'unique' must be drug money.[25]

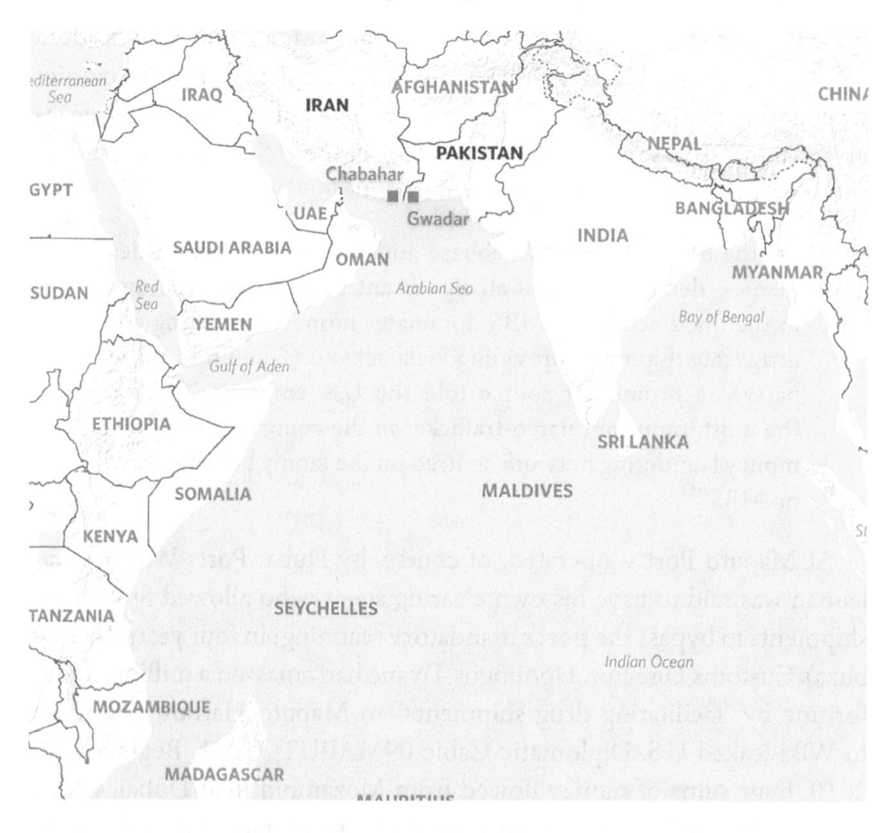

Between Pakistan and Mozambique: The Heroin Coast

The Heroin Coast and Natural Gas

Historically many heroin boats put in at the port of Mocimboa da Praia[26] in Mozambique's Cabo Delgado province, where in 2010 gas was struck by Texas oil company Anadarko.

25 J. Hanlon, "The Uberization of Mozambique's Heroin Trade," London School of Economics, 2018

26 Global Initiative Web site, April-May 2020

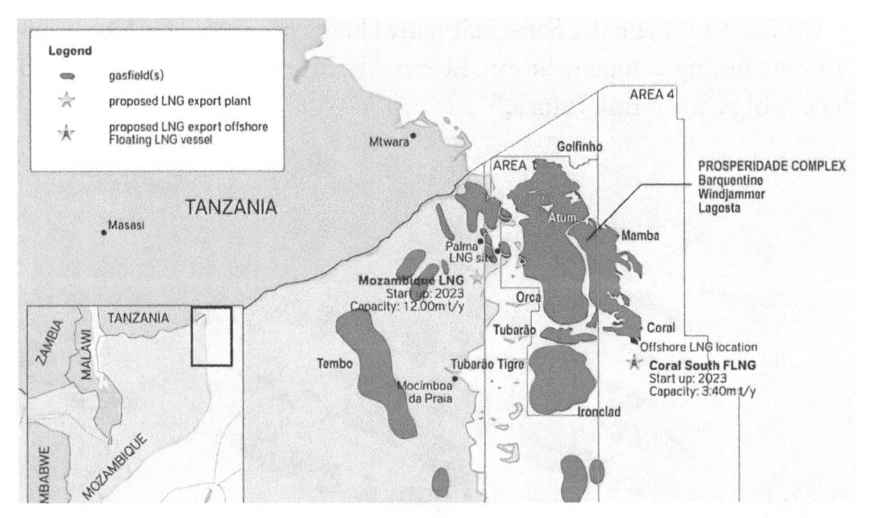

Both onshore and offshore, Mocimboa da Praia (center in the above graphic) is flanked on three sides by Western-claimed gas deposits. Problematically for Big Oil, however, heroin *dhows* from Pakistan routinely sail directly over these offshore deposits, preparing to put in at Mocimboa da Praia. Mozambique lacked boats to shut down this traffic (this highly likely is because Guebuza was interested in maintaining the drug traffic – it benefited him). But once any gas started pumping, publicity would reveal that Mocimboa da Praia, site of a growing Western "Gas City," was also once a major heroin port enriching the Mozambican officials that granted the gas permits in the first place. Such a revelation was clearly intolerable. The only solution was a patrol fleet owned by the kind of Westerners that were skilled in preventing publicity – namely, spies. And this fleet is the one Iskandar Safa built that was acquired by a do Rosario company that sold the boats to skilled American spy Erik Prince.

Payment to Safa for these boats came from a fake "loan to Mozambique" from Credit Suisse, arranged by Safa lieutenant Boustani, represented as money with which Mozambique would buy a fleet of "tuna boats" from Safa. As part of this scam, Safa employee Jean Boustani arranged a $2 billion "loan" from Credit Suisse ostensibly for Mozambique to use in buying tuna boats from Privinvest.[27]

But the $2 billion went directly from Credit Suisse to Privinvest, much of it then disappearing into covert ops by do Rosario's SISE[28] and into bribes and kickbacks for officials at Credit Suisse and in Mozambique.[29]

27 Reuters, October 16, 2019

28 Financial Times.com Web site, February 28, 2019

29 Organized Crime and Corruption Reporting Project, July 24, 2019. Also cf. "Don't pay that 'debt', Mozambique," IndependentOnline (IOL) Web site, April 23, 2020. Safa has claimed mon-

Do Rosario made the Safa-built patrol boats property of a Mozambican state fishing company he ran. Eventually, his company, Ematum, sold the patrol boats to Erik Prince.[30]

Ematum fast patrol boats unused and on the hard at the Port of Maputo African Ports & Ships Maritime News, June 20, 2018.

As of this writing, Mozambique still does not operate a coastal-security fleet, and Western gas operations are stalled, in the midst of a three-year-old civil war called an insurgency. Western nations are pressuring for large-scale military occupation of the Cabo Delgado gas field, onshore and offshore, in the Ruvuma River Basin. U.S. Green Beret troops are already there, as "trainers," as of early 2021.[31] This is reminiscent of Vietnam.

Gas City

Conveniently for Western oil companies, the border-area civil war has caused more than 115,000 coastal residents to abandon Cabo Delgado.[32]This depopulation is crucial because the companies plan for the area a "Gas City" on 7,000 hectares,[33] an unfarmable, unfishable, unlivable swath of roads, pipes, parking lots, and company buildings that will

ey to Mozambique's Filipe Nyusi was campaign contributions. Cf. "Ex-Credit Suisse Banker Admits Kickbacks in Africa Scandal," FinNews.com Web site, July 22, 2019

30 Reuters, December 13, 2017 A great majority of the Ematum boats from Safa were patrol boats and interceptors (as pictured above, this page). A minority were fishing boats.

31 *New York Times*, April 7, 2021

32 ReliefWeb Internet site, April 8, 2020. The "pressure-socialist-Tanzania" angle is not fanciful; *the BBC has referred to "reports that some militants are Swahili speakers, suggesting they could have connections with Tanzania."*

33 Plataforma Media Web site, November 14, 2019

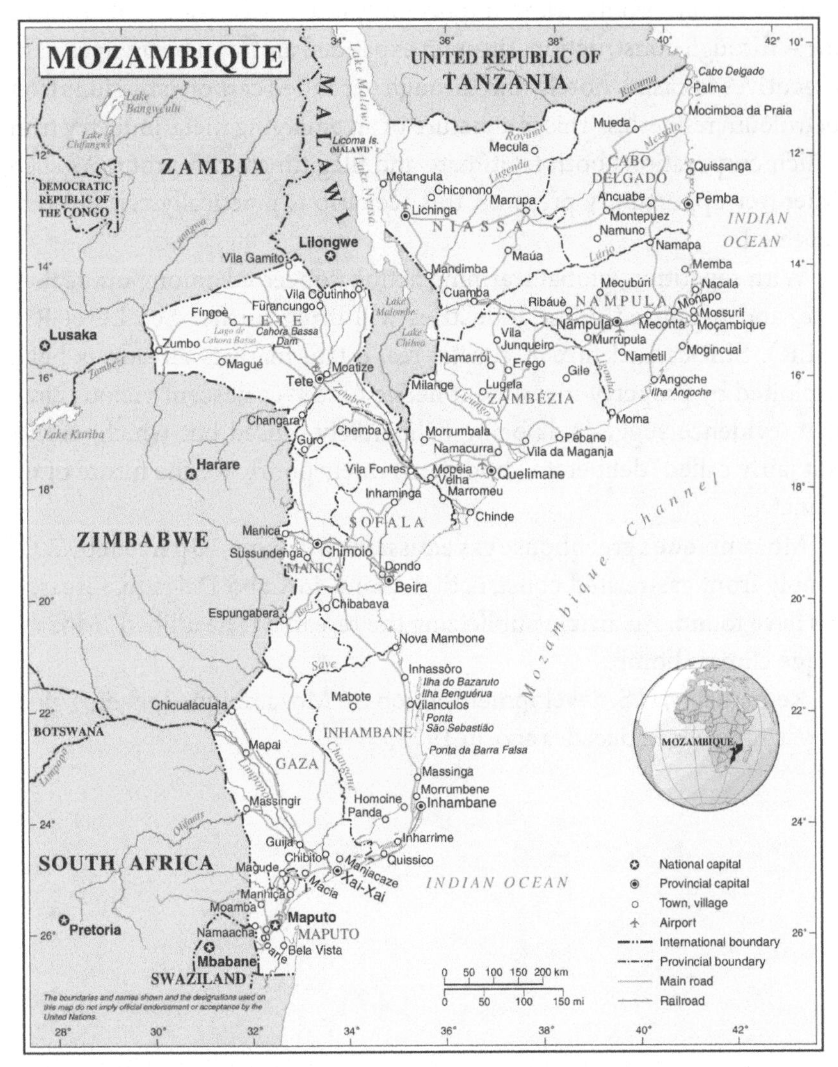

The Ruvuma River, near top in graphic, separates Mozambique's Cabo Delgado province and neighboring Tanzania.

eclipse shoreline for miles south from the Tanzanian border on the Ruvuma River. This is a shoreline to which we will return in the final chapter of this book. Half of the Ruvuma Basin gas field, which is enormous, lies in Tanzania, a socialist country, where as in Mozambique Western corporations aim to extract petroleum and minerals – essentially extending northward the MozamScam described in this chapter.

Mozambique is set for extension southward of Gas City, a coming megalopolis of sprawl from Mocimboa da Praia-to-Maputo of oil infrastructure and resort/casinos. This will profit gangsters who build casinos to launder heroin profits. More importantly, it will accelerate global warming – through construction, through expanded air travel by invitees of oil executives to casino hotels, and through expanded carbon emissions from petroleum refineries. This is a picture of an emerging meta-industry from which corporations both legitimate and illegitimate can profit. As such, wherever opportunity presents, this scenario is practically certain to be repeated.

With repetition, global warming action concerted among oil, real-estate, and organized crime industries will bring further Sea Level Rise (SLR). SLR-caused flux in coastal real-estate markets already is being exploited by property-investment hedge funds – a nascent vicious circle that, evidence suggests, is being deliberately caused but which, even if not fairly called "deliberate," bodes extremely poorly for the future of the planet.

Mozambique's greenhouse gas emissions will jump 10 percent by 2022 simply from gas-related construction, alone, in Cabo Delgado, researchers have found. An article publicizing the research is headlined "Mozambique climate bomb."[1]

Remarkably, U.S. development actions in Mozambique speeding global warming began decades ago, in 1975.

Chapter Two

Mozambique Concedes Coastal Lands To Americans: Jim Blanchard And Greg Carr

American oilmen's promises to "lift Mozambicans out of poverty" began decades ago when Texan James Blanchard proposed not oil development but tourism. He got a large government land concession in 1996, to develop tourism on Mozambique's "Elephant Coast." The back story is notable.

James U. Blanchard III

In 1975, the U.S. secretly supported fighters against a newly independent, and socialist, government of Mozambique. These fighters were recruited by RENAMO (Resistência Nacional Moçambicana), a group formed and based originally in neighboring Rhodesia, now Zimbabwe.

A Renamo recruiting station, likely in Rhodesia, now Zimbabwe

Part of this secret U.S. support for Renamo was American Jim Blanchard, who with a former CIA deputy director's son-in-law, hiked in with supplies of walkie-talkies for these fighters, who were based on

Mount Gorongosa. Throughout the conflict, Renamo fighters deforested much of the mountainside cutting trails, slaughtered thousands of elephants for ivory to buy weapons, and hunted antelope and zebra for food.

Mount Gorongosa

While this was going on, to support Renamo Jim Blanchard sold to fellow Texas oilmen gold coins emblazoned "Renamo," used the proceeds to buy hand-held radios, and in person distributed the radios in Mozambique to Renamo soldiers. Years later, Blanchard's son would be arrested in Zimbabwe/Rhodesia, original home of Renamo, for possession of "weapons of war."[1]

Twenty years of attrition and bribery by Western forces softened Mozambican resistance, and in 1996, dealing with a post-socialist Mozambican government, Blanchard got a land concession. Also that year, Texas company Enron got an oil concession in Mozambique. It is important to note that at this time, between Texas interests (including Cheney's Halliburton) and African officials, bribery was the order of the day.[2] As such, it is fairly likely that a single tranche of Texas bribes eased both deals – the Blanchard land concession and the Enron oil concession.

Blanchard was granted more than half a million acres in coastal Mozambique,[3] including the Maputo Elephant Reserve. As Africa Confidential reported on April 25, 1997,

> The once-socialist government of Mozambique has made what may be the world's most extreme privatization agreement. It is handing over the development of a region as large as Israel to a

1 BootsnAll Web sit, August 1999

2 Cheney's Halliburton/KBR bribed Nigerian officials between 1995 and 2004. NBC News, December 7, 2010.

3 InterPress news agency, November 22, 1999

> private firm owned by Texan billionaire James Ulysses Blanchard III, a former supporter of the Resistência Nacional Mozambican (Renamo). The agreement, approved by the Council of Ministers last December, was meant to be secret, but Africa Confidential has seen the text.

This secret deal privatized the Maputo Elephant Reserve.

On March 13, 1999, Blanchard transferred $60,000 to the Maputo Elephant Reserve fund administered by government officials, likely a camouflaged bribe in that three months later, Ponto Dobela in the Elephant Reserve was dubbed to become a coal seaport to handle shipments by South African interests, under agreement by Mozambican and South African officials.[4]

So, what?

So, Jim Blanchard had become a front for a deep international intrigue aimed at seizing Mozambican coastal lands for fossil fuels extraction and for real-estate development. As such, Blanchard was in over his head. One week after he transferred the $60,000, Blanchard was dead.

He died of a drug injection, in a Louisiana motel room. Local papers noted the coroner found a mix of cocaine, benzodiazepine, and "an unknown opiate" in Blanchard's body.[5] His wife said he had no history with narcotics. Another coroner told me that "unknown" is used when a drug isn't among those on which a preferred commercial lab offers testing services, frequently a very recently synthesized drug. Such drugs include analogs of Fentanyl popular with organized-crime dealers and intelligence agents. National and worldwide media omitted reference to the fatal opiate, reporting cause of death for Blanchard as "heart attack or stroke."

Greg C. Carr

On Mount Gorongosa, the Blanchard-supported Renamo fighters eventually killed 95 percent of the zebras, for food, and killed 95 percent of the elephants, in trading ivory for weapons.

They cut thousands of trees. In 2004, precisely because Renamo fighters had depredated the area, Idahoan Greg Carr – with U.S. State Department connections – won a large land concession at Mount Gorongosa, essentially the right to run the 1.1-million-acre park. Carr promised to restore wildlife. He also promised to expand tourist accommodations that

4 Recall from Chapter 1 that at this time spies from South Africa and the U.S. likely had begun their plotting to exploit Mozambican gas resources in Cabo Delgado).

5 Stephanie Hanes, 2017 - *Nature*

would employ Mozambicans[6] and to build clinics and schools. More than 20 articles lauding the project appeared in print between 2009 and 2019, including the *New York Times*' "In Mozambique, a Living Laboratory for Nature's Renewal," nearly all of them written by journalists who spent comparatively little time, or none at all, at Gorongosa.

In contrast, according to journalist and author Stephanie Hanes, who traveled with Carr and spent months in Gorongosa interviewing locals, the Restoration Project failed to recognize that Mount Gorongosa is sacred to locals, possessed of a spirit that is angry at the mountain's desecration over recent decades – by Western-funded anticommunist Renamo soldiers. Many of these Gorongosans viewed Carr's project as extending a long-running spiritual malaise, known as "The Disorder," that traditional tribal gods had visited upon Africa. Carr's Western philanthropy showed sympathy for people without material goods, but it lacked empathy for their spiritual stance. Hanes recounts how some responses by Carr to her questions aimed to dissuade her from publishing her book, with perceived veiled threats against her future career.[7]

With Carr's promise of clinics and schools, in 2009 the U.S. State Department backed Carr strongly, announcing that the department's U.S.AID was committing millions of dollars for

> ... encouraging sustainable economic activities inside the (Mount Gorongosa) park – and around its buffer zone.[8,9]

This was, as it happens, just at the time the MozamScam was gelling. Antonio Carlos do Rosario and Armando Guebuza had risen to power in Mozambican government, and, remarkably, their government deliberately was not seizing any heroin.[10]

Carr eventually distanced himself from this ruthless power game; in 2008 he essentially turned his Gorongosa concession over to U.S.AID.[11]

Jim Blanchard was not so fortunate.

---

6 *Smithsonian Magazine*, May 2007

7 Hanes, "White Man's Game: Saving Animals, Rebuilding Eden, and Other Myths of Conservation in Africa," Metropolitan Books, 2017

8 U.S.AID Web site This U.S.-established "buffer zone" became home for former Renamo fighters, under peace accords.

9 Associated Press, August 1, 2019

10 www.jjcornish.com For five years, Jean Jacques Cornish edited *The Star & SA Times International*, a newspaper for South African expatriates. Also cf. J. Hanlon, "The Uberization of Mozambique's Heroin Trade," London School of Economics, 2018

11 U.S.AID Bureau for Economic Growth, 2018.

Chapter Three

Nigeria and Angola: Cheney

Jim Ballard's case gives a glimpse of rough boys, playing for keeps, inhabiting or at least fringing the complex of extractive and development industries that is bent on global warming – a milieu in which ends, money and power, justify means ranging from murdering a man to murdering a planet.

This amoral utilitarianism appears in the1995-through-2004 bribing in Nigeria by Dick Cheney's Halliburton/KBR,[1] in the 1997 Project for a New American Century. Project members agreed a desired type of pretext for a U.S. oil hegemony worldwide was "A new Pearl Harbor," and the events of September 11, 2001 provided that pretext. Scholar Peter Dale Scott has said Cheney needs to be questioned under oath about foreknowledge of 9/11[2] The amoral utilitarianism also appears in Ballard's death in 1999, in Cheney's secret 2001 Energy Task Force, and in the case of the Boko Haram movement in Nigeria.

Boko Haram

Amoral utilitarianism characterizes Dubai, a power base for the spies-and-illicit-money MozamScam. The record suggests some of this illicit money went to support Boko Haram, and that contrary to mainstream news accounts of Boko Haram, much of this support money came from Westerners. Indeed, it appears the Boko Haram cannot be adequately understood except in context of the utter dishonesty of Nigerian officials that was nurtured heavily by bribes from Western entities such as Cheney's Halliburton/KBR. This dishonesty began with a massive U.S. arms sale to Nigeria, then festered while government officials embezzled and sold much of this weaponry, and finally broke into the mainstream news with the revelation that some of these arms were being sold to Boko Haram, by the Nigerian military.[3]

The weapons sale, initially approved by President Barack Obama, was delayed upon reports the Nigerian military was detaining and torturing

1 PBS, February 12, 2009. Also cf. Panafrican News Agency (Dakar) May 1, 1999.

2 ; Cf. Chapters 13-16 of my *Not Exactly the CIA: A Revised History of Modern American Disasters*, Trine Day, 2020-21.

3 Associated Press, September 4, 2016

thousands of Nigerian Muslims. Almost as soon as the U.S. arms arrived in Nigeria, a Dubai connection turned up along with the revelation that the military was selling some of the arms to Boko Haram. Nigerian Lt. Gen. Tukur Buratai was charged with buying, using embezzled cash, two properties in Dubai, worth $1.5 million.[4] Dubai bank records were so sketchy that prosecutors couldn't obtain firm evidence that the cash had been embezzled.

Corroborating a Dubai connection, reporting for the *Daily Trust* newspaper in the capital of Abuja, a Nigerian journalist wrote,[5]

> Around 2017 I learnt of a dramatic arrest of a Dubai-based Nigerian unlicensed currency exchange operator.
>
> (He contracted) to transfer the money from Nigeria to the Dubai-based currency exchange operator. who takes his share and transfers the balance to the fraudster in leaving the currency exchange operator the actual receiver of the money according to the official records who therefore ends up being traced and arrested in Dubai.
>
> After going through it, I discovered, among other things, that the currency exchange operator in question was actually one of a 6-man network of accomplices convicted for Boko Haram funding transactions at least since 2015, nearly all initiated by one Alhaji Sa'idu based in Nigeria, called an undercover Boko Haram member in Nigeria.
>
> Wondering whether Nigerian authorities are aware of this case, I checked with a Nigerian diplomat friend of mine here in the UAE.

The Nigerian government maintains a diplomatic mission in the UAE, which is manned by supposedly trained diplomats as well as National Intelligence Agency (NIA) operatives, who are particularly responsible for preventing foreign-linked threats against Nigeria and its interests.

None of this fits the standard Western account of the Boko Haram war as simply Islamic terrorists versus a legitimate Nigerian democratic government. In the context of rampant Western bribery in Nigeria, however, it was easy for these seemingly counter-intuitive events to occur.

Cheney's KBR bribes[6] bought not only access to petroleum fields, but also the use of Nigerian police and military as security guards for oil operations, which would draw protestors. Thus, the bribes bought murder

4 Voice of America, September 4,2016
5 *Daily Trust*, Abuja, August 21, 2020, published on the AllAfrica Web site
6 Nigerian courts indicted Cheney on bribery charges; he settled the indictment out of court.

and mayhem. In September 1997, on the orders of Halliburton/KBR contracting for Chevron, officers of Nigeria's Mobile Police shot and killed protestor Gidikumo Sule at an oil-flow station in the city of Warri.[7]

Halliburton's bribery practice began when the Nigerian government launched ambitious plans to build the Bonny Island Natural Liquefied Gas Project. With Bonny Island, Nigeria's global-warming carbon emissions rose 22 percent between 1994 and 2005.[8]

To the south, in Angola, the same thing happened. Halliburton was there, the dos Santos dynastic government was dishonest (banking offshore millions in Western bribes), and by 2018 an Angolan minister who knew too much, as had Jim Ballard, was dead by apparent foul play, in a hotel room – in Mozambique. This was Adao do Nascimento.

Just before Cheney assembled his secret Energy Task Force, his Halliburton was locking up an Angolan deal with patriarch Jose Eduardo Dos Santos.

Africa Confidential, January 12, 2001, wrote,

> Halliburton, the former company of Vice-President-elect Richard Cheney, has $350 million in loan guarantees by the United States government through the Exim Bank for work in Angola where it has six offices and works with Chevron, on whose board sits the new National Security Advisor Condoleezza Rice.[9]

Also around this time, dos Santos was speaking of giving up his presidency. But, for some reason, dos Santos changed his mind. He then won a further election in 2003, and then held onto the office until 2017.

Also around the time of Halliburton's arrival, privatization was under way in Angolan higher education, a corruption-riddled system in which dos Santos's nephew – Adao do Nascimento – was notably instrumental, rising in the Angolan education department to full Minister of Education by 2005. Just then, KBR was up for a separate contract award, and Nascimento's vote on the Council of Ministers' was necessary to KBR's getting that contract.[10]

Under this regime, Angola's global-warming carbon emissions fully sextupled[11] *while Angola's population merely doubled.*

---

7 Macrotrends Web site

8 MacroTrends Web site

9 Cheney met in August 2001 with Kenyan President Daniel Arap-Moi for a half hour, then said a brief hello to President Bush.

10 RigZone Web site, April 25, 2005

11 Between 1994 and 2014. Macrotrends Web site

One of dos Santos's final moves was to fire Nascimento, by then known for "unreliability." Twenty months later, Nascimento was dead, in Mozambique. Maputo authorities found traumatic head injuries."[12]

12 ClubOfMozambique Web site, December 19, 2018. Autopsy found signs of heart attack.

Chapter Four

Colombian Cocaine

Maputo, in addition to being a through-point for heroin trafficking, is distribution point for cocaine that arrives in great quantities from South America. The luxury hotel where Adao do Nascimento died likely bore some connection to organized crime.

The Mozambican agency investigating do Nascimento's death, SERNIC, collaborates with Carlos do Rosario's SISE in cases where an organized crime-government connection might emerge.[1] This chapter describes how intelligence operatives have aided the organized-crime industry, part of the complex of industries whose concerted activities are accelerating global warming.

Recall that Mozambican stock exchange chairman Jussub Nurmamade said, "What makes Mozambique 'unique' must be drug money." As a U.S. Embassy official wrote in 2009,

> The primary route for cocaine is by air to Maputo from Brazil via Johannesburg, Lisbon, or Luanda (Angola). The drug traffickers routinely bribe Mozambique police, immigration and customs officials in order to get the drugs into the country.[2]

Brazilian Gilberto Aparecido dos Santos[3] and his organized-crime company PCC[4] long has been chief exporter to Maputo of South American cocaine. Dos Santos, nicknamed "Fuminho,"arrested in Maputo by the U.S. DEA in April 2020,[5] uses a franchise system. This franchise system is modeled after Colombia's Medellin cocaine cartel – dues from member drug dealers pay lawyers, bribe police and prison guards, and buy weapons for a protection service.

As it happens, the record shows the CIA and the Mossad helped start the Medellin cartel – as a black market for weapons that the U.S. couldn't

1 ClubOfMozambique Web site February 14, 2020

2 U.S. Charge d'Affaires in Maputo Todd Chapman, in leaked U.S. Diplomatic Cable "09MAPUTO1291.

3 No relation, I believe, to Angola's Jose Eduardo dos Santos.

4 Primeiro Comando da Capital ("First Command of the Capital).

5 Agence France Press, April 14, 2020

sell legally to "sanctioned" nations, particularly Panama.[6] The record shows the Mossad assisted here, by brokering such weapons under Israeli stamp, to Panama's Manuel Noriega.[7] As we shall see, some weapons were left over, and spies arranged to sell them to Colombians who trafficked cocaine.

It was in 1975 that demand for cocaine accelerated worldwide. As a result, economies changed in South America, with farmers migrating to government-deforested lands in Peru's Huallaga Valley. They planted coca. Dealers flew leaves to Colombia for refinement into cocaine. A $50,000 "tax" was levied by former head of Peruvian intelligence Vladimiro Montesinos on each coca-carrying plane he allowed to leave the Huallaga Valley.[8]

By organizing the Medellin cartel with the bait of weapons to arm a cartel "security force," the CIA, too, was able to monitor and skim Huallaga Valley drug proceedings.

The key to the CIA's gaining this ability was a particular U.S. military-intelligence operation called "Watch Tower."

Beginning in 1975, U.S. Army Air Force Operation "Watch Tower" monitored and assisted cocaine flights from Colombia to Panama.[9] In Panama, for a "special security force," ruler Manuel Noriega wanted special weapons he couldn't get legally from the U.S.," such as small missiles.

According to CIA operative Trenton Parker, Colombia's Medellin cartel began when, at a meeting organized by the CIA at the Hotel International Medellin,[10] some 200 dealers in attendance agreed to pay $35,000 apiece for services of a security force armed to repel police attacks – armed, as research indicates, with security weaponry – including shoulder-fired heat-seeking missiles – originally supplied by the Mossad for Noriega's security unit.[11]

U.S. military special-operations veteran and mercenary trainer Frank Camper writes that at a Panama hotel,

> I was being requested to provide shoulder-fired heat-seeking antiaircraft (from the Mossad's supplying of Noriega) for the Medellin cartel.[12]

---

6 Officially, former U.S. ally Manuel Noriega was sanctioned, as a drug runner.

7 Peter Dale Scott, *Cocaine Politics: Drugs, Armies, and the CIA in Central America*, University of California Press, 1998

8 BBC, April 6, 2001

9 Rodney Stich, *Drugging America: A Trojan Horse*, Diablo Western Press, 1999

10 O. Villar, U.S. Narcocolonialism? Colombian Cocaine and Twenty-First Century Imperialism, "Research in Political Economy, 2007 This large meeting had been preceded by an earlier one attended only by Colombia's 20 top dealers.

11 Ibid. to 7

12 *Orlando Sentinel*, July 17, 1988

A Noriega agent invited Camper to a gathering in a Panama hotel in 1981. On June 5, 1989, citing intelligence sources *The Spotlight* reported that the Medellin cartel of Colombia was, in reality, "an Israeli-directed organization which nets Israel billions of dollars every year in illegal drug profits."

On June 4, 1990, *The Spotlight* cited an inquiry by the Government of Antigua (Antigua had been accused of receiving arms shipments destined illegally for Colombian end users). When this was leaked in September 1989 to international media outlets, Rafael Eitan, a former chief of staff of the Israeli Army, told the Israeli press frankly:

> Someday, perhaps, if it's decided that the stories can be told, you'll see that [the government of Israel] has been involved in acts (in Central America) that are a thousand times more dirty than anything going on in Colombia. These things were decided by the government, in cabinet meetings. As long as the government decides to do something, something that the national interest demanded, then it is legitimate.

The CIA, with right-wingers, and organized crime also were acting in concert in Central America.

In April 2006, a DC-9 carrying nearly 6 tons of cocaine and co-owned by a committee appointee of Congressional Majority Leader Tom Delay was busted at an airport in Ciudad del Carmen, Campeche, Mexico bearing an official-looking seal painted on its side reading SKY WAY AIRCRAFT, PROTECTION OF AMERICA'S SKIES, around an image of a federal eagle.[13]

In a stunt probably attributable to Fuminho dos Santos and his PCC, organized criminals with ties to Latin America's palm-oil industry dissolved a ton of cocaine in a palm-oil shipment bound from Colombia to Belgium.[14] The record shows cocaine ties abound in the palm oil industry,[15] which is booming. As such, palm-oilsters connected to Colombian coke are among the leaders in the complex of crime-tied industries whose concerted actions are accelerating global warming.

Researchers have found that each hectare of rainforest converted to oil palm monoculture creates 174 tons of carbon emissions.[16] In just two years of a Colombian palm-oil boom, between 2014 and 2016 the nation's carbon emissions per capita rose 10 percent while its population rose less than 3 percent.

---

13 *Mad Cow Morning News*, April 17, 2006

14 Europol Press Release, May 31, 2011

15 So do ties to U.S. intelligence.

16 Mongabay Web site, July 5, 2018

United Fruit Shipping-Line Advertisement

Chapter Five

Organized Crime, Spies, and The Oil of Palm

In Colombia, a boom began in 2014 for a second kind of Big Oil industry, palm oil, which, like the petroleum industry, encompasses large corporations and large production operations, including refineries. And, the palm-oil industry is populated and fringed on several continents by organized criminals and spies.

Just weeks after the South Africa Public Investment Company curiously invested $80 million in the S&S Palm Oil Refinery, under construction in the Nacala, Mozambique, Container Park, fire destroyed a warehouse next to the nascent palm-oil plant.

Like the container park, the warehouse was owned by a member of the dishonest Maputo business elite around Armando Guebuza, discussed in Chapter 1. Investigators found the fire at the Nacala Industrial Park, home to the S&S refinery, destroyed a shipment of motorcycles, some with imported heroin in their tanks. [1] The S&S refinery was under construction at the time, and after that fire, it never opened for business. As such the property, owned by Guebuza crony Momade Rassul, never refined any palm oil – but with its stink of organized-crime – created a large scandal in the South African government, which had invested in the property.[2]

Around the globe, from scores of palm oil refinery effluent ponds, methane and carbon dioxide are continually rising, contributors to global warming along with petroleum, coal, and organized crime.

For decades in Latin America, a concert of actions has served to accelerate global warming. These are the concerted actions of United Fruit, Halliburton, Drummond coal, Glencore traders, and organized-crime members connected to members of the elite such as banker Jaime Rosenthal and palm-oil magnate Miguel Facusse.

---

1 J. Hanlon, "The Uberization of Mozambique's Heroin Trade," Working Paper Series, No. 18-190, London School of Economics and Political Science.

2 ClubMozambique Web site, October 5, 2018

Organized crime and palm oil

Organized crime was connected to palm oil's very beginning as a commodity. After that, organized crime was connected to palm oil's becoming entrenched as a profitable industry, and after that, organized crime was connected to palm oil's continuing profitability into the 21st century.

Around 1895, European ships bound to the Caribbean from Africa carried both abducted African people and the African oil-palm tree. At this, American industry giant United Fruit stepped in.

An industry expert writes,

> The early history of the oil palm in Central America is largely the history of the crop in the United Fruit Company.[3]

The Port of New Orleans docks were controlled by a dockworkers' union controlled in turn by the Sicilian Mafia's Macheca family.[4]

Joseph Macheca

As such, in order to gain access to the Port of New Orleans, United in a 1901 merger with a Macheca-family company[5] bought the Macheca's shipping line. United carried fruit in the mob ships to New Orleans docks for six full years before building its own banana boats in 1907.[6]

Also that year, machine gun-toting organized crime forced abdication of the Honduran president.

Samuel Zemurray, who would later become president of United Fruit, hired well-armed thugs including Guy "Machine Gun" Molony and Lee Christmas, sailing with them from New Orleans to the Honduran coast where the novel Molony machine gun defeated a Honduran military response, the sitting president stepped down, and in a new election Zemurray's Honduran friend Manuel Bonilla returned to power.

A later right-wing political elite in both Honduras and Colombia in the 1930s took to cocaine smuggling and in the 1980s to employing death squads to steal peasant lands for coca and palm plantations while granting permits to United Fruit, Drummond Coal, and Glencore. This elite included Miguel Facusse and Jaime Rosenthal, and United, Drummond

---

3 OilPalmBreeding blogspot, November 29, 2009

4 Thomas Hunt and Martha Macheca Sheldon, *Deep Water*, Createspace Independent *Publishing* Platform, 2010.

5 Peter Dale Scott, *Deep Politics and the Death of JFK*, University of California Press, 1996.

6 For decades to come, these New Orleans docks remained under Mafia control, primarily under Carlos Marcello, one of the most powerful of U.S. Mafia bosses.

and Glencore cooperated with them in both drug smuggling and use of death squads.

Honduras took to cocaine smuggling and palm oil planting in the 1930s because its economy was crashing from banana blight[7] and just at a time when United Fruit and Standard Fruit each was getting 500 hectares of land from the government for every kilometer of railroad track they could build. In 1940 Standard Fruit's founder Lucca Vaccaro was indicted for extortion – along with all other members of the Port of New Orleans governing board.[8]

In 1949, when Dick Cheney was a youngster, the Halliburton company later to be Cheney's patented a combination of palm oil and naphthalenic acid, "na-palm."[9]

In Korea, U.S. bombers dropped 32,557 tons of Halliburton napalm, largely against civilians, deemed "communist." This, of course, was U.S. state terrorism, and a U.S. Army general spelled this out in 1962, writing,

> A concerted effort should be made now to select personnel for clandestine training (to) execute paramilitary terrorist activities against known communist proponents. It should be backed (covertly) by the United States.[10]

Expressing this policy in Colombia and Honduras were death squad paramilitaries who doubled as guards for palm plantations.[11] At some plantations, including those held by palm-oil magnate Miguel Facusse guarding meant protecting cargoes of cocaine landing on plantation airstrips.[12]

Coca for this cocaine was grown largely in Peru's Huallaga Valley, where Halliburton was working and where U.S.-funded deforestation[13] had allowed migrant peasant farmers to plant coca fields and allowed Halliburton to build roads to prospect for oil. In 1990 Halliburton's presence in the Huallaga Valley was intrusive enough that its oil-surveying camp

---

7 Ibid. to 4.

8 TreeHugger.com Web site, November 9, 2018.

9 Kansas Geological Survey, Public Information Circular (PIC) 32. Eventually, Halliburton would use napalm in fracking.

10 Secret supplement to a February 1962 U.S. Army Special Warfare Center report by General William Yarborough to the Joint Chiefs of Staff. Wikipedia

11 E.g., the notorious Autodefensas Unidas de *Colombia* (AUC)

12 ForeignPolicy Web site December 11, 2019. Also cf. a March 19, 2004 cable from the U.S. Embassy in Honduras: "a law enforcement source provided information that the aircraft successfully landed March 14 on the private property of Miguel Facusse, a prominent Honduran.""

13 Under the "Upper Huallaga Special Project," a five-year, $26.5 million program funded mainly by the U.S. Agency for International Development, which works closely with the CIA. Wikipedia

was attacked by leftist guerrillas.[14] From every cartel pilot flying coca out of the Huallaga Valley toward Colombian refineries, $50,000 went to ex-head of Peruvian intelligence Vladimiro Montesinos.[15]

After packaging on palm-oil plantation airstrips in Honduras, most of this Colombian cocaine went north in flights to the U.S., generating millions in profits laundered through Honduran banker Jaime Rosenthal as loans to buy palm-harvesting equipment.[16]

Between 1970 and 1975, as much of the world took to cocaine, the Colombian coca-planting industry expanded.

Just after this, indicted briber Garry Drummond[17] of Drummond Coal was finding coal in Colombia, and revolutionaries were gaining ground nearby. So, (following U.S. policy) Drummond covertly financed "paramilitary terrorist activities" by the mercenary group AUC in a years-long spree of violence against dispossessed peasant farmers. At this time, an Exxon-Colombia joint venture was opening Cerrejon coal mine, the world's largest open-pit coal mine. This mine would be acquired in 2016 by multinational Glencore, a Central American palm-oil trader as well as coal miner. Also at this time, right-wing business sectors in Honduras were hiring mercenaries from Colombia who later would form the Autodefensas Unidas de Colombia (AUC).[18]

It is highly likely Drummond paid his AUC mercenaries with CIA money skimmed from the drug trade – in 1988 Drummond managers met former CIA agent James Lee Adkins in Miami and in 1995, Drummond Coal hired Adkins as its security advisor in Colombia.[19]

Soon, during just the years 1992 and 1993, representing Drummond and other coal companies the National Coal Association and petroleum interests spent 1.8 million on ad agency Burson Marsteller, arguing that *"The role of greenhouse gases in climate change is not well understood."* And while the global-waming complex paid Burson Marsteller with the right hand, *with the left it was funding Colombian mercenaries* in AUC when AUC was responsible for scores of deaths in the 1990s.[20] This utilitarian amorality by the global-warming complex is remarkable, but ultimately, it is unsurprising.

14 *Los Angeles Times*, December 13, 1990

15 BBC, April 6, 2001; El Pais, September 29, 2016

16 *U.S. v. Rosenthal*, Filed 12/08/17 U.S. District Court Southern District of New York

17 Indicted but not convicted of bribing Alabama officials.

18 Cf. "The dark side of coal," PAXColombiaReports Web site, June 25, 2014.

19 PublicIntegrity.org Web site, July 22, 2014

20 At Mapiripan and at Cano Jabon, Colombia, in July 1997 and May 1998, respectively. ColombiaReports Web site, February 17, 2020

AUC funds came from Drummond and from a host of palm-oil traders, 19 of which were indicted in 2012 by the Colombian Prosecutor General's office for funding paramilitaries.[21] At this time, funding for AUC and for Honduran paramilitaries also came from a team of some 20 palm oil corporations.[22] Gustavo Duncan, security analyst at University of the Andes, wrote

> Palm was a perfect way to consolidate their militarized social control over a territory and invest capital accumulated from drugs (cocaine) into a profitable business.[23]

AUC funds also came from a subsidiary of Glencore,[24] which is a major trader of palm oil, a world-wide miner of coal, and a major shipper out of the Dubai Ports World-run Maputo Port in Mozambique. AUC funding also came from United Fruit (Chiquita), while AUC used Chiquita ships, launched from Chiquita ports, to move cocaine in the thousands of pounds; according to one source, employees of Chiquita knew about the drugs.[25]

Thus, with help from CIA agent Landry who worked for Garry Drummond's coal company in Colombia, the coca industry helped fund the CIA's Contra war, as the CIA supplied an air fleet and an airstrip on the Honduras ranch of dishonest banker Jaime Rosenthal's crony Ramon Matta Ballesteros (a known major cocaine dealer). Then, Ballesteros pilots landed U.S. arms near the Nicaraguan border for the Contras and carried U.S.-bound Colombian cocaine to northern Mexico. In 1986, records show, the State Department paid $185,924.25 to Matta Ballesteros' airline SETCO.

In this spy-and-gangster ridden milieu, Halliburton had made palm into a solid industry, by patenting napalm. *The reason Halliburton patented napalm was for use in fracking, a notorious global warmer.*

---

21 ColombiaReports.com Web site, January 15, 2015.

22 Ibid.

23 The Intercept Web site December 23, 2017

24 "The Octopus in the Cathedral of Salt," *University of Virginia Quarterly Review*, Fall 2007. It is likely that much of this funding was CIA-skimmed drug money using the companies as fronts

25 Melissa Sartore, Ranker.com Web site, September 23, 2021

Chapter Six

The Deepwater Horizon Blowout and Spill, and Coastal Real Estate

In 1996, fracking advocate Dick Cheney took the reins at Halliburton just as the global-warming issue was arising. Big Oil, blamed for the warming in published evidence, suppressed the issue by spending millions on counter-arguments. Companies hired scientists soon known as "The Merchants of Doubt," who had worked as deniers for the tobacco industry, to say that lack of conclusive proof meant, "It's OK to do nothing until conclusive proof exists."

But then, in a notably serious move, in 1997 the BP oil company broke ranks with its fellows – BP admitted Big Oil was accelerating global warming.

BP's CEO John Browne said,

> The time to consider the policy dimensions of climate change is not when the link between greenhouse gases and climate change is conclusively proven, but when the possibility cannot be discounted and is taken seriously by the society of which we are part. We in BP have reached that point.[1]

Then, around 2000 BP said its initials stood for "Beyond Petroleum." To anyone in Big Oil who has spent large money denying petroleum's global warming, and who happens to be easily offended, this remark amounts to an insult. Would it be passed over? The record suggests it was not passed over.

Dick Cheney is a man known for using great power vindictively. He is also known as a master of the long game in arranging American access to oil – cf. the 18-year Iraq-Afghanistan wars he began. As such, there's some likelihood Cheney and his secret 2001 Energy Task Force cronies planned a payback for BP – something to make BP hurt for its leaving the group of petroleum-climate-change deniers. BP did in fact suffer – after the 2010 Deepwater Horizon blowout. BP was vulnerable then because it

1 E.Malone. "Debating Climate Change: Pathways Through Argument to Agreement - p75," *Earthscan*, 2009. [Archived .png on file, DeSmog Web site]

already had a poor environmental record; blowout-disaster fines threw a $64-billion punch to BP's pocketbook.

Less known is that the Deepwater disaster also profited Texas oil companies, including Halliburton and Murphy Oil. Texas oil interests show a poor record on caring about gargantuan oil spills. In *1958, Amoco joined Gulf in a Mozambique oil concession, in which Gulf and Amoco drilled the Pande Wells #3 and #4. The #4 well blew out and flowed uncontrollably at an estimated 1 billion cubic feet per day for 400 days before the blowout was capped.*

The year 1996, when Big Oil was committing to a hard climate-denial battle, was a time of great political flux in Mexico, from which Big Oil and the U.S. real-estate investment industry would benefit – as these related industries set up Mexico for a profitable exploitation, one which would accelerate global warming.

The Texas-Pemex Scam: Oil, Los Zetas, and U.S. Spies

Mexico's petroleum company is called Pemex. Importantly, as of 2005, Texas refineries preferred, and bought, Pemex's "natural-gas condensate" over Texas's own crude oil. Because Pemex -gas condensate often contains few contaminants and is easily refined into high-value oil products, it generally competes directly and favorably with Texaslight crude oil in downstream oil markets. [2] Then in 2006, in the midst of a worldwide oil-price spike, Pemex stopped shipping its gas condensate to Texas refineries – a problem. Texas oil outfits solved this by dealing with organized crime – buying Pemex condensate from the Los Zetas organized-crime cartel.

The Zetas had developed an elaborate system of skimming this petroleum from Pemex's border-area pipelines and, using tanker trucks disguised as Pemex vehicles, carrying the oil across the Texas border. From 2005 to 2007, DEA and DHS agents were wiretapping members of the Zetas dealing cocaine in Dallas, and likely in that way learned that Zetas were skimming Pemex oil and selling it in Texas."[3]

Remarkably, Texas oil companies began their illicit-oil buys from the Zetas just as the Zetas were joining with an Italian mafia, 'Ndrangheta ("brave men"), to traffic cocaine from Mexico to Italy[4] (more later on 'Ndrangheta).

2 Court proceedings, Pemex v. numerous Texas-oil defendants, June 2, 2011 Courthouse News Service.

3 DEA, July 23, 2009. Also cf. MarketWatch.com Web site, August 22, 2009.

4 This became evident during the U.S.'s Operation "Project Reckoning." InsightCrime Web site, December 12, 2016

About a year later, Mexican authorities formally notified U.S. officials of the Texas-Pemex scam.[5] It became evident that a Texas outfit called Continental Fuels was buying and storing stolen Pemex product for distribution to refineries.

At this news, Vice President Cheney did nothing (both understandably and for a bad reason). Why? Almost certainly because Cheney was connected to Continental Fuels.

Only after Cheney left office was a U.S. legal case brought – against Cheney's former press advance man Josh Crescenzi, a purchasing officer for Continental Fuels.[6]

After Cheney and before Continental, Crescenzi's employer was Cheney's one-time security advisor – oilman and lobbyist Stephen Payne.[7] In Payne's network, Crescenzi also worked indirectly for early Cheney mentor Frank Carlucci Jr., deputy CIA director from 1978 to 1981. Cheney's first job in government was under agency Director Frank Carlucci in the Office of Economic Opportunity, 1969-1970. Carlucci would become a covert CIA agent in the Congo, conspire there with Belgian operatives to assassinate Patrice Lumumba, and later become Deputy Director of the CIA. In the Texas-Pemex case, his protégé Crescenzi assisted prosecutors and was never punished.

As such, the record of the Pemex-Texas caper strongly suggests that government officials assisted the collusion among top oilmen and members of Mexican organized crime. And for any such collusion, liaisons are necessary – men who could walk easily and be respected in both the oil world and the crime world. These, of course, are "spies" – either on staff, or fired, or retired, etc. from the myriad of intelligence organizations.[8] There is evidence for this happening in Pemex-Texas.

Just as Crescenzi began to buy stolen oil for Continental Fuels in 2007, Continental was taken over by shady operator Kamal Abdallah and his soon-to-be valueless company, UPDA (Universal Property Development & Acquisition). U.S. Securities Exchange Commission investigators found UPDA to be a "shell" company and Abdallah to be a fraudster and jailed him.

On June 9, 2010, SEC revoked UPDA's securities registration because soon after acquiring Continental, UPDA stopped filing required reports with the SEC, leading the agency to say in the case,

5 MySanAntonio Oct 9, 2009
6 And other defendants. *Houston Chronicle*, October 10, 2009
7 At Payne's oil company, Worldwide Strategy Energy.
8 See my book, *Not Exactly the CIA*, Trine Day, 2020-21.

> Many publicly traded companies that fail to file on a timely basis are "shell companies" and, as such, attractive vehicles for fraudulent stock manipulation schemes.[9]

As such, Kamal Abdallah has the profile of a man useful initially to spies – here, as a front owner for Continental – but then made a fall guy.

It's well worth asking how Crescenzi, talented and credentialed into the U.S. power elite, could wind up at a place like Continental, off the beaten track and soon to be judged a valueless shell company. In any case, at Continental Crescenzi taped his phone calls with oil sellers and buyers,[10] and then gave the evidence-packed tapes to Homeland Security's ICE border patrol, which, with this evidence made numerous arrests.[11]

As such, the record suggests Crescenzi was a plant – in some covert operation designed by his former associates Carlucci, Payne, and Cheney to fully control Pemex – and its oil deposits on the Mexican Gulf coast. What promotes Texas oil interest promotes "national security." Eight months after fall guy Kamal Abdallah was arrested, Deepwater Horizon blew.

The Spill

The 2010 Deepwater Horizon oil-rig blowout in the Gulf of Mexico was quickly deemed a criminal matter because 11 people died in the blowout. Manslaughter charges were filed against two BP employees, later dropped. *Halliburton pleaded guilty to destroying evidence around faulty cement it knowingly supplied for the BP oil rig.* [12]

In addition, in nine separate safety tests (by Chevron) of the same cement (prepared from a Halliburton recipe), the Halliburton cement failed all nine times. It never passed. The cement supplied by Halliburton was so bad it only lasted one day before the blowout came.[13]

During ex-Halliburton CEO Dick Cheney's secret Energy Task Force meetings in 2001, among self-serving laws he allowed oil executives to draft[14] one law said no "acoustic switch" was required on offshore oil rigs – a device that seals a well permanently if a blowout occurs.

Attorney Mike Papantonio said,

---

9 e-Smart Techs., Inc., 57 S.E.C. at 968-69 n.14. 2

10 *Houston Chronicle* August 2, 2011

11 InSightCrime Web site December 12, 2016

12 Department of Justice, September 19, 2013

13 BP, "Deepwater Horizon Accident Investigation Report," September 8, 2010.

14 Including the notorious "Halliburton Loophole" under which Congress exempted fracking from compliance with the U.S. Clean Water Act.

> An acoustic switch would have prevented this (Deepwater Horizon) catastrophe – it's a fail-safe that shuts the flow of oil off at the source – they ... are required in off-shore drilling platforms in most of the world ... except for the United States. This was one of the new deregulations devised by Dick Cheney.[15]

As a result, with Cheney's effective permission, the Deepwater Horizon rig had *no acoustic blowout-prevent switch.*

And remarkably, Deepwater partner Transocean had *deactivated the rig alarm and had left it off.*[16]

Inescapably, suspicion must arise at this point. That is, given the facts – no prevent switch and no alarm – *it was only a matter of time* before Deepwater Horizon blew; *a spill was certain to happen.* As far as I know, this simple albeit dismal conclusion from the public record, as cited, has not previously been publicized.

However, it certainly poses a chilling question – "Was the blowout disaster intentional?"

Here, what can separate supreme indifference from active interest? Suggestive of active interest is a profit motive for the entire "landsters/oilsters/banksters/gangsters/securisters" interest group. The record shows profits accrued quickly after the disaster, for example, to Texas oil companies including Halliburton and even to BP. Global warming accelerated.

Of the 4.9 million barrels of oil spilled by Deepwater Horizon, corporate "cleanup" crews burned off a total of 245,000 barrels, releasing more than 1 million pounds of black carbon (soot),[17]*) roughly equal to that released by all ships sailing the Gulf in a 9-week period. Per unit of mass, black carbon causes 460 to1,500 times more global warming than carbon dioxide.*

And, as we shall examine, Gulf real estate dropped in price, even in the U.S., while the Mexican Gulf coast was thrown into chaos. A pliant U.S. press did not report on this chaos, apparently because both the chaos and silence about it were in the interest of the powerful landster/oilster/bankster/gangster/securister crowd.

Coverup

For years after the Deepwater spill, U.S. media wrote not a word about oil damage to Mexico's Gulf coast. Instead, they wrote BP's claim – in hindsight, preposterous – to have data showing oil had drifted south

15 DeSmog Web site Saturday, December 17, 2011.

16 Ostensibly so no false alarm could disturb employees' sleep. Denver Post, July 23, 2010; BBC, July 24, 2010

17 *Science Daily*, September 21, 2011

to Brownsville, Texas, but, there, had stopped – and had not drifted the less than three miles further south to Matamoros, Mexico. Oil companies deliberately concealed data in order to mislead the public, and the press helped, even though the existence of false data was referred to in public Texas court proceedings, in the case of State of Veracruz of Mexico v. BP, Halliburton, et al.

The court record reads in part,

> Scientific studies by oceanographers from the Universidad Nacional Autonoma de Mexico demonstrate and establish that the oil slick and the underwater plume of oil released into the Gulf will also spread westward with the wind and currents towards the coastline of Veracruz in approximately October and November of this year (2010).

In the case, attorneys argued,

> The oil spewed into the Gulf's waters at a flow rate and in amounts that BP refused and failed to acknowledge and/or accurately report. From the outset, BP continuously reported to the regulatory agencies and others that the flow rate and amounts spewing forth from its well were but a tiny fraction of the actual rates and amounts. BP's actions in this regard had the predictable effect of lulling regulatory authorities into a more passive response and were attempts to preserve its already tarnished public image and avoid financial penalties. Such conduct and false information disseminated to the public and others had the effect of lulling others into falsely believing the environmental impact was not as severe as they would have otherwise believed and causing them to defer taking appropriate preventative and mitigating action.[18]

However, a Texas judge ignored evidence of BP concealment intending to mislead and ruled instead on the basis that Veracruz lacked sufficient "proprietary interest" in Mexican land to base a damage claim. Thus, the case was over quickly and received no mainstream U.S. reportage. I was able to find no reference to the case in the *Houston* (Halliburton home town) *Chronicle*, the *Dallas Morning News*, or the *San Antonio News-Express* (the clearly important case was featured in publications to which attorneys may pay for subscriptions).

18 Case 5:10-cv-00761-OLG Document 2 Filed September 16, 2010 in Texas federal court;. Republic of Mexico v. BP, Halliburton et al

It was a British newspaper, the *Guardian,* that more than five years after the blowout first reported BP oil pollution of Mexican coastal communities. On December 11, 2015, the paper reported a class-action suit representing claims by Mexican coastline dwellers. This was after BP had spent money eventually totaling $500 million to create the Gulf of Mexico Research Initiative to "study the effect of hydrocarbon releases on the environment and public health," an outfit that requested independent researchers to submit their spill data; this made the data dependent on BP for release, and evidently BP suppressed independently gathered spill data, such as the following:

> The oil reached Mexican shores on 30 April (2010, 10 days after the blowout). Hundreds of communities which rely on fishing and tourism in the worst-affected states of Tamaulipas, Veracruz, Tabasco, Campeche, Yucatán and Quintana Roo have seen their livelihoods plummet. The damage is ongoing, according to the claim.

An oyster fishery more than 600 miles south of Brownsville was wiped out by spilled oil; in Veracruz, shrimp and bass fisher families between these cities saw their incomes decline as much as 90 percent. The Saladero, Veracruz fishing cooperative in 2010 took 11,663 kg of shrimp, 36k g of bass and 281,125 kg of oysters. In 2019, the co-op registered only 1,000kg of shrimp, 20kg of bass, and no oysters.[19]

Scientists 10 years later, after it was too late, corroborated 100-percent the claims that BP concealed data to mislead the public. An international scholarly journal, Science Advances, *reported that the Deepwater spill likely was nearly 30 percent larger than generally publicized and that U.S. researchers deliberately withheld data from the public at the time of the spill.* [20]

> "Although the spill took place almost 10 years ago, *some of the data wasn't released until recently* (in 2020)," the Web publication Verge said in covering the Science Advances article.

Verge cited study authors and Ira Leifer, a researcher at University of California Santa Barbara. Leifer was a member of a technical group tasked with putting together official estimates of the flow of oil from the spill, and he says the study confirms what a decade of observations, medical records, and anecdotes had implied: "The impact of this spill was larger than generally publicized," he said.

19 *Guardian* April 19, 2020.
20 February 12, 2020

During the 10-year breather from scrutiny that the oil companies got from the Texas judge's ruling in *Veracruz v. BP & Halliburton,* they finagled access to Mexico's publicly owned oil, in a scheme called the "reform" of the public oil utility, Pemex.

The misleading label "energy reform" refers to opening publicly owned Mexican oil to private investment.

In the "reform" of Pemex, President Enrique Pena Nieto granted BP, Halliburton, and Texas's Murphy Oil lucrative contracts – even though in a lawsuit Murphy had admitted buying Pemex condensate found to be stolen by Zetas,[21] as such, likely Halliburton, BP, and Murphy could have done no better if they had written the legislation themselves called "Energy Reform."

And effectively, they had, in that a leading member of their oil-and-real-estate interest group, Citibank/Citigroup, previously had laundered cartel millions in bribes to Mexican officials[22] and in 2013 succeeded in privatizing a Mexican real-estate-investment trust that specialized in hotels.

Damaged by BP oil, Mexico's Gulf coast, like America's, was no longer primarily home to folk whose work was fishing. It was for hoteliers and oilmen.

As a Veracruz local told the *Guardian,*

> Now, (people) are forced to migrate – to find factory work in *maquilas* in faraway cities.

Real Estate, Venture Capitalists, Cartels

All around the Gulf coast, prices dropped after the BP spill for waterfront real estate. Developers bought up land cheaply, and Americans profited handsomely – especially in Alabama and the nearby Florida Panhandle. Between 2010 and 2015, in a five-county stretch, population grew by an average of more than 10 percent, many of the arrivals buying newly built resort condos.[23] The surge was so pronounced that the coastal city of Fairhope, Alabama, imposed a moratorium on new construction.

BP settlement money fueled this boom, which involved numerous casinos and thus may have attracted organized crime. In Biloxi, Mississippi, BP money helped build a stadium for MGM Resorts International – to house ball games next door to MGM's Beau Rivage Casino. The Beau Ri-

21 BuzzFeed Web site, September 28, 2018. Murphy Oil lawyers admitted the buys but denied the company knew the oil was stolen.

22 Raul Salinas Gortari, for whom Citi laundered $100 million in cocaine money in the mid-1990s. Raulis the brother of Carlos Salinas Gortari, Mexican president at the time.

23 Condo Owner Web site January 17, 2017

vage was built by Steve Wynn, a reported front for the Genovese organized-crime family (Scotland Yard found this mob connection).[24]

Reported mob associate Stanley Ho and his family have had a long-running joint venture with MGM Resorts International.

Much of the post-BP-spill construction of casinos, condos, and hotels on the Gulf Coast was done by Birmingham-based developers. Among Birmingham's top 20 land developers and partnering on at least one occasion in the Alabama/Panhandle area are Ron Durham and BL Harbert International (BLHI).

BLHI, a major construction player worldwide, earned U.S. taxpayer money in 2015 building a new U.S. Embassy in Matamoros, Mexico, a site of considerable oil damage from the BP spill. Not long after, Durham, partnering with real-estate and construction firm Skanska, earned millions in BP settlement money for work on Alabama's Lodge at Gulf Shores State Park project.

Skanska and The Lodge at Gulf State Park

Skanska has an odd record.

Perhaps the best job Skanska worked recently in its home country of Sweden was building a new home for Sweden's spy agency, the SSS, in 2020. But in a matter adjudged likely to involve organized crime, Skanska deliberately supplied faulty cement for the Mater Dei Hospital project in Malta in 1996. When the hospital concrete failed, it emerged that Skanska had a zero-liability clause in its contract.[25]

In an inquiry in Malta, retired Judge Philip Sciberas wrote of Skanska,

> It is evident that such a defect (in the hospital concrete) could not be the result of a genuine mistake or failure of oversight but must have been the result of a concerted effort from which the contractor, suppliers, and possible third parties benefited.

It is is virtually certain that by "third parties" Judge Sciberas is referring to mobsters either Sicilian or Calabrian, the latter being the 'Ndrangheta that during the Texas-Pemex stolen-oil scheme in 2006 was partnering with the Los Zetas cartel to traffic cocaine to Italy.

This lucrative cocaine trade in February 2010 allowed the Zetas to wrest control of northeastern Mexico from the previously more powerful Gulf Cartel.[26]

24 Cf. "Steve Wynn, the Scotland Yard Report," www.stevemiller4lasvegas.com Wynn's customary source of funding was junk bonds from his friend, later-convicted fraudster Michael Milken.

25 *Malta Today*, May 25, 2015

26 Wikipedia

Two months later, BP's Deepwater Horizon blew. Two months after that, in June 2010 came billions in BP settlement money to buy/redevelop oil-degraded, cheapened real estate. Toward sharing that money pot, the record suggests, effective collusion tightened among Big Real Estate, spies, and organized crime.

A Skanska manager in Ecuador admitted that Skanska hired mercenaries to keep locals away from its zone of operation and used bribes to avoid legal trouble[27] (this of course mirrors actions by the palm oil-cocaine industrial complex in Colombia and Honduras, where the United Fruit/Drummond Coal "Octopus" funded organized terrorist-mercenary groups such as AUC after Garry Drummond hired a CIA agent as consultant).

Just at this time, while Skanska was building the headquarters for Sweden's national spy agency,[28] it was *buying data from spies to use in blacklisting* thousands of environmental activists and unionists from working in the construction industry.[29] Skanska in 2015 got a BP-settlement-funded contract to work on Alabama's Lodge at Gulf Shores State Park.

Borderlands

Before that, Skanska was making money in 2012 along a drug route established by Los Zetas crossing the Rio Grande from Piedras Negras in northeastern Mexico to Eagle Pass, Texas. The Skanska work site was the Lucky Eagle Casino on the reservation of the Kickapoo Band of Traditional Indians, within yards of the Rio Grande. There, Skanska managed construction of a new hotel and an expanded casino space. Also there, cross-river smugglers employed by Los Zetas moved from riverbank canebrakes to vehicles waiting in the casino's crowded parking area – whose drivers could then blend safely into the casino's exit traffic and, next, disappear in the nearly lawless town of Eagle Pass.[30]

As is somewhat well-known, the CIA routinely monitors and skims off established drug routes. Good examples from the early 1980s include CIA skimming of a Mexico-to-Los Angeles cocaine route in the 1980s to buy weapons for the Contras[31] and CIA monitoring of a Sinai-to-Egypt opium and heroin route.[32] The record strongly suggests the Zetas' Eagle Pass cocaine route long has been monitored by the CIA.

27 UpsideDownWorld Web site, November 6, 2007

28 Skanska Web site, July 1, 2010

29 *Belfast Telegraph*, March 27, 2015; Wikipedia

30 U.S. Customs agents testified to this in court. *San Antonio Current*, August 3, 2016

31 (Cf. Gary Webb, *Dark Alliance: The CIA, the Contras, and the Crack Cocaine Explosion*, Seven Stories Press, 1998

32 Cf. my *Not Exactly the CIA: A Revised History of Modern American Disasters*, Trine Day, 2020-21.

For example, Eagle Pass is a town greatly influenced by the notorious Wackenhut company, a CIA asset, which built a detention center there.[33] The Border Patrol and Wackenhut are the main employers in Eagle Pass. Out of Eagle Pass in 1990, virtually certainly with CIA knowledge, Wackenhut trucked a mustard-gas ingredient bound illegally for Iraq's Saddam Hussein.[34] This was at a time when Wackenhut was joint-venturing with American Indian Tribe Cabazon Band of Mission Indians on the tribal reservation, a location outside U.S. legal jurisdiction (except on crimes of violence). Dick Cheney and other U.S. officials cited mustard-gassing by the Iraqi regime as justification for the 2003 U.S. Iraq invasion.

Cross-Border Shale Play, Mexico Ripe for Fracking

Eagle Pass is surrounded by the massive Eagle Ford shale-gas play. Mexico has roughly the same holding in the Eagle Ford play as does Texas. Dick Cheney's Halliburton dominates the Texas side.

In 2013, just as Mexico's Texas-oil-brokered "energy reform" privatization was set to begin, Halliburton opened a new office especially to work Eagle Ford.

> "The company's South Texas operations have grown in stride with the Eagle Ford," a Halliburton spokesman wrote.

It was fairly clear Halliburton in concert with "reform" privatization of Mexican oil reserves planned to stride further, into Mexico's portion of the Eagle Ford play.

When Mexican President Andres Manuel Lopez Obrador was elected in 2018, he said he would ban fracking. But Lopez Obrador wavered, and today, fracking is at an impasse, with still no law either for or against. U.S. investors are awaiting more "administrative reform" that would sway legislators.

Well, not exactly waiting. In August 2020, Halliburton-offshoot Recon Africa announced permits to frack in Africa's Kavango Basin. In a shale play said to "dwarf" the Eagle Ford, an Eagle Ford veteran, former Halliburton contractor Nick Steinsberger, will run the project for Halliburton's former vice president for Eagle Ford Scot Evans, now CEO of Recon.

The project spans the border of Namibia and Botswana, where, the record shows, public officials initially lied about fracking in the Kavango Basin.

---

33 SourceWatch Web site, December 25, 2019

34 John Connolly, "Inside the Shadow CIA," *Spy Magazine*, September 1992.

Chapter Seven

Fracking and Global Warming

In Kavango Basin in September 2019, the U.S. Department of Homeland Security had intervened,

> ...host(ing) a law enforcement training workshop in Botswana in August 2019 concerning the Kavango Zambezi Trans-Frontier Conservation Area.[1]

This allowed Botswanan officials to appear concerned and proactive about conserving wildlife and habitat in the Kavango-Zambezi area, when in fact they had allowed fracking for years in Kavango.[2] Less than a year later, in an apparent quid pro quo, from the same officials Recon had its permit, but in continuing dishonesty, Botswanan officials denied that Recon would use fracking at Kavango.[3]

Recon, on the other hand, was never shy about stating its fracking plans. CEO Scot Evans said in 2020,

> (Fracking is) a technique now utilized in all commercial shale plays worldwide. (Kavango project supervisor) Nick Steinsberger is the pioneer of "slick water fracs."[4]

Such frankness, close to arrogance, signals blithe indifference about global warming – fracking emits to atmospheric up to 35 times more greenhouse gas than does conventional drilling.[5]

Under Dick Cheney's administration came a frenzy to build U.S. frack sites, aided by the Cheney Energy Task Force and the Halliburton Loop-

---

1 Press release, September 19, 2019, U.S. Immigration and Customs Enforcement. This area aims to conserve wildlife ranging from elephants to rosewood trees.

2 Photojournalist Jeff Barbee, "Massive Secret Data Leak Exposes Fracking in Botswana," AllianceEarth Web site July 25, 2017.The government of Botswana first denied these facts, but we had in-person testimony on camera from the drillers on the ground, and both film and photography of fracking waste water pits, fracking pumps, sand trucks, and the (Kalahiri Energy) company's own public reports online that showed fracking projects that had been going since at least 2005," Barbee wrote. "After numerous denials through the press the Botswana government finally admitted it was happening. We broke the news with a story in the UK Guardian and our first short film was released on the investigative news programme Carte Blanche in 2013."

3 FairPlanet.com Web site October 16, 2020

4 Reconnaissance Energy Africa Ltd. ,June 25, 2020,

5 *Science Daily* July 25,2019

hole for fracking it got amended to the U.S. Clean Water Act. Astoundingly between 2000 and 2015, Halliburton and other Oilster companies opened approximately 277,000 new U.S. frack sites[6] (fracking now accounts for 67 percent of U.S. gas production).

Oil-sands territory currently being fracked already emits more greenhouse gas than do the nations of New Zealand and Kenya combined.[7] If all the oil in just those sands could be burned – with evolving fracking measures – another 240 billion metric tons of carbon would reach the atmosphere – enough by itself to raise global temperature by .03 degrees Celsius.[8]

Fracking Deforestation

Texas Big Oil is committed to fracking, around the planet. Forests, the planet's carbon trap, bind carbon molecules (in photosynthesis), preventing them from warming the planet by reaching the atmosphere.

We have seen that each frack well emits more carbon than a conventional rig. But, in a double whammy, each frack well deforests more, too – creating an unfortunate equation: *for each net acre of forest lost, each subsequent emission is more likely to reach the atmosphere.*

A conservatively estimated 6.5 acres of trees are cut just for one forestland fracking pad[9] (other estimates are much higher,[10] and this average size seems to be increasing).[11] This is all in addition to acreage cut for access roads, for storage tanks, and for wastewater ponds.

As such, it is unsurprising that at least as early as December 2016, scientists warned fracking "could be a critical factor driving climate change."[12]

Remarkably, just three months after this warning was published, in March 2017 fracking boomed in Colombia – with at least 43 new con-

6 U.S. Energy Information Administration, March 15, 2016; cited by Ballotpedia

7 *National Geographic* July 9, 2020

8 Ibid.

9 Moran, M.. "Habitat Loss and Modification Due to Gas Development in the Fayetteville Shale". Environmental Management, 2015.

10 The Nature Conservancy estimates that a fracking operation, what with the well area, roads, equipment storage ground, and wastewater pits, disturbs 12 hectares (approximately 30 acres) of land in Pennsylvania's Marcellus Shale area, between 2012 and 2018 the size of a typical fracking site increased by 30 percent to more than 6 acres.

11 In Pennsylvania's Marcellus Shale area, between 2012 and 2018 the size of a typical fracking site increased by 30 percent.

12 Qingmin Meng, "The impacts of fracking on the environment: A total environmental study paradigm," Department of Geosciences, Mississippi State University, 2016 Unfortunately, as we shall see in coming chapters, to particular interests – the oilsters , the landsters, and the banksters – this finding could either be unimportant or serve as a useful business tip – to increase fracking just because doing so will increase acceleration of global warming and thus depopulate more coastal lands faster, for profitable exploitation.

cessions for Drummond, ExxonMobil, ConocoPhillips, and other U.S. companies.[13]

Flaring

Flaring is the practice of burning gas that is deemed uneconomical to collect and sell. This amounts to prodigious waste – university researchers found that Texas has flared about as much gas annually as all of its residential users consume.[14]

Always producing a black smoke, this habitual burning is prodigiously global-warming, as well. Black smoke concentrates more carbon particles than light-colored smoke.

Flames and black smoke from gas flaring

In Texas, even excluding the Eagle Ford play, flaring from the Permian Basin alone contributes approximately 1 percent of man-made carbon emissions globally.

And, industry under-reports flaring emissions to regulators.

Comparing satellite-based data with figures that industry rerported for 2012-2015 to Texas regulators, researchers found,

> In total, the volumes reported in the state database were only around half of what the satellite observed."

Similar studies found similar discrepancies for shale regions in New Mexico and North Dakota.

When Texas A&M researchers turned up the Texas discrepancies in 2020, others were showing, piece by piece, that *concerning global-warming effects, practically all data, reported from the entire oil industry, have been underestimates.*

13 Colombia National Hydrocarbons Agency, World at 1 Degree Centigrade Web site, April 7, 2017

14 Gunnar Schade, Kate Willyard, Texas A&M Today Web site, August 3, 2020

Chapter Eight

Outdated Industry Estimates That Allow Accelerated Global Warming

Under-reporting is the rule for all industry-caused global-warming effects. This applies equally to specific events like oil spills and to general patterns. Industry habitually has relied on inadequate methods of data gathering, and as a result has reported practically nothing correctly.

To gather oil-spill data on Deepwater Horizon, BP used a standard, accepted industry procedure. *However, scientists eventually showed that the spill was 30 percent larger than BP reported* (see "The Spill" section of Chapter Six). This is ominous – called into question are all oil-spill data reported by industry.

Also unreliable have been industry reports of how much greenhouse gas they've emitted. Here, too, industry data-collection simply follows an accepted industry standard, and recent science has showed *this industry-standard method underestimates carbon emissions.*

In 2018, Stanford University researchers found that planet-wide, oil-industry emissions are 42 per cent higher than the industry reported. This constitutes a full 5 percent of global emissions. In comparison, total global emissions from aviation is roughly 2.7 percent.[1] Also that year, Environmental Defense Fund researchers found *methane leakage at Pennsylvania frack sites five times higher than industry reported,*[2] and in 2019 an Environment Canada researcher found greenhouse-gas emissions in Canada to be 64 percent higher[3] than industry estimated by "using internationally recommended methods." This is a full 17 metric tons higher.

A researcher said,

> The quantity 17 metric tons per year is similar to the greenhouse-gas emissions from a metropolitan area the size of Seattle or Toronto.

---

1 M. Masnadi et al, "Global carbon intensity of crude oil production," Science, August 31, 2018.

2 EDF.org Web site, February 15, 2018

3 J. Liggio et al, "Measured Canadian oil sands CO_2 emissions are higher than estimates made using internationally recommended methods," Nature Communications April 3, 2019.

And, researchers warned,

> "Given the similarity in reporting methods across the entire Oil&-Gas sector, these results suggest that (all) Oil&Gas CO_2 emissions... data may be more uncertain than previously considered.

This is uncomplicated, if ominous. The public record shows plainly that *the oil industry has supplied us with, and has asked us to rely on, estimates of industry global warming that have been far too low.*

This means that historically, society has lacked any data-gathering method to show accurately global warming by the oil-industry. It is unfortunately clear that until this changes, we lack firm basis for crafting realistic oil regulations.

Real-Estate Industry, Too

Likewise, unsurprisingly, the real estate industry has offered underestimates of global warming, which depreciates values of coastal real estate. according to published scientific research, coastal properties already had depreciated by 15 percent due to Sea Level Rise caused by global warming by 2020. Researchers had found,

> (Coastal) properties are trading at a 14.7-percent discount (as) sophisticated investors actively discount properties based on Sea-Level-Rise exposure. [4]

However, major media effectively denied this type of finding by researchers. Instead, relying on trusted real-estate industry sources, the *Tampa Bay Times* reported on January 17, 2020,

> Due to sea rise, your Florida coastal home could lose 15 percent of its value by 2030 (not 2020, as published research had already shown.)

This discrepancy reflects that the real-estate industry ignored this research or carelessly was unaware of it. Additionally, the industry passed on this misconception to prominent media. Whether deliberate or just careless, this kind of falsehood is important because purchase of depreciated property gives investor/developers extra capital to spend building profitable raised or sea-walled condo communities. Precisely when the coasts are claimed by Sea Level Rise depends, of course, on the degree to which global warming continues to be accelerated. The record strongly

4 A. Bernstein et al, "Disaster on the Horizon: The Price Effect of Sea Level Rise," *Journal of Fnancial Economics,* 2019. "Underwater," a study from the Union of Concerned Scientists, predicts 300,000 U.S. coastal properties will flood "before 2050." In any case,

suggests the latter depends on whether the cohort can be reined in that includes oilsters, banksters, gangsters, and real-estate investors – landsters,.

Recent research has shown that more coastal land will be under water sooner than predicted on a standard model accepted in the real-estate industry.5 Many investors in coastal real estate, especially overseas, consult GIS Planning, a service from the *Financial Times*,[6] which also is now the standard tool for any government Economic Development Department wanting to attract commercial investment. This standard tool for Big Real Estate fails to predict how fast sea level rise will inundate a particular property, researchers found.

Why? The tool mistakes treetops and building tops for surface land, researchers found.

> SRTM7 models the elevation of upper surfaces and not bare earth terrain. It thus suffers from *large error* with a positive bias when used to represent terrain elevations. This is especially true in densely vegetated and in densely populated areas." This *degree of error* leads to *large underestimates* of Extreme Coastal Water Level exposure, and *exceeds projected sea-level rise this century under almost any scenario*.[8]

Developed by NASA, the erroneous STRM model was acquired quickly and widely by planners and investors overseas – because it is free on the Internet. This widely spread source of error concealed the true situation from planners overseas, who invited in foreign land investors.

Each developer who has thus – improperly in a sense – acquired lands just inland from villages now owns part of the coming "new coast."

In Yucatan, Mexico, the Gulf port city of Progreso was projected to see its sea level rise by a foot as soon as 2030, according to the most dire 2016 projections by Climate Central.[9]However, the chief scientist at Climate Central has recently co-authored the new, 2019 research, according to which the dire 2016 projection wasn't dire enough. Even sooner than thought, Mexico's Gulf coast will see villages of fishers and farmers displaced by Sea Level Rise, making way for Big Real Estate interests through the FibraInn investment trust.

5 S. Kulp, B. Strauss,. "New elevation data triple estimates of global vulnerability to sea-level rise and coastal flooding," Nature Communications, 2019

6 GIS Planning's ZoomProspector Enterprise and Intelligence Components software are in use around the world. In January 2017, Financial Times, perhaps the top investor publication, acquired GIS planning.

7 NASA's "Shuttle Radar Topography Mission" geographic-information system

8 Researchers corrected for error using "neural network" (artificial intelligence) technology.

9 *Forbes*, January 8, 2016

Reporting on the new research, the *New York Times* on October 29, 2019 published the following graphics:

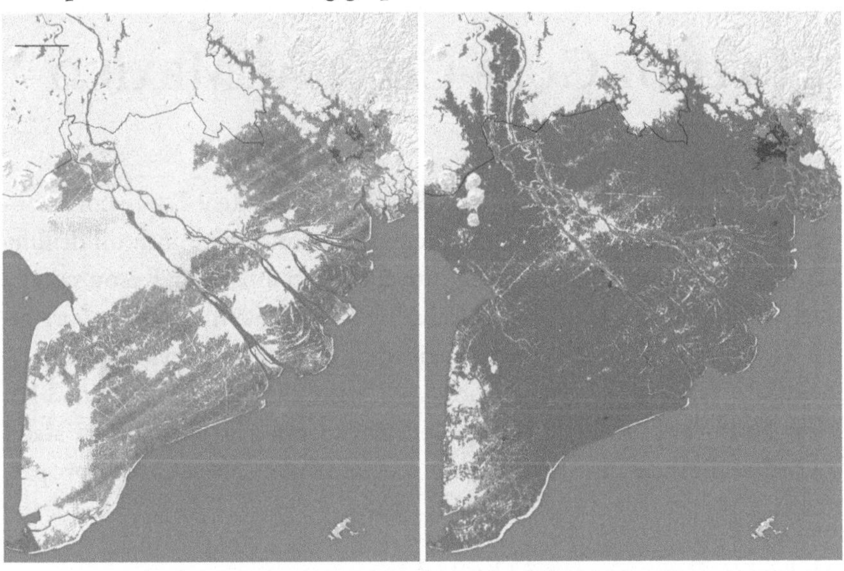

Land underwater at high tide, Cambodia
Old projection for 2050 New projection for 2050

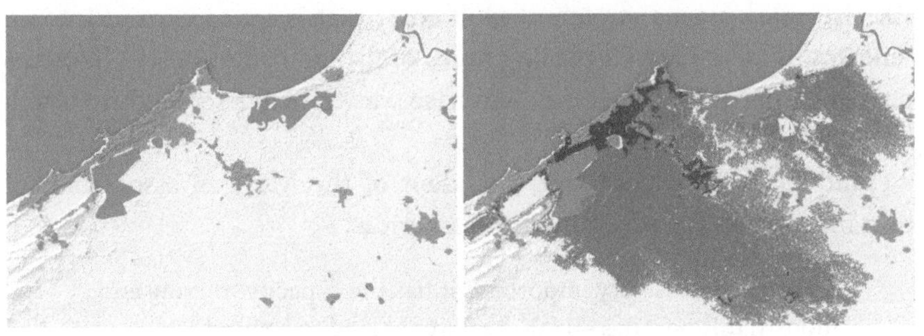

Land underwater at high tide, Egypt
Old projection for 2050 New projection for 2050

When society has tried to check on how much global-warming damage the system of oilsters, landsters, and securisters is causing, the only data it has found available are now-discredited estimates from industries in the cabal.

Chapter Nine

Fate of Gulf-Coastal Mexico

In Mexico, before 2014 neither Big Oil nor Big Real Estate could operate as they pleased. The Mexican state controlled both oil drilling and real-estate investment to restrict foreign intrusion. Resourcefully overcoming this obstacle, Big Oil and Big Real Estate managed to "privatize" (open to U.S. capital) both Mexican oil and Mexican real estate.

While Texas oil companies brokered privatization of state-owned Pemex, New York-based Citigroup[1] led privatization of FibraInn, a state-owned real-estate investment trust specializing in hotel/casino properties.[2]

Altamira, Reynosa, and Ciudad Juarez

The first hotel purchase by FibraInn, in 2014, was in Altamira, on a BP-oil-damaged stretch of Gulf coast that is set to expand into an "enormous" Gas City, displacing miles of fishing communities. Trustees proudly emphasized the hotel purchase was to better serve visitors to a local Halliburton office.

Julio Martínez Hernández, president of the Mexican association of port infrastructure (AMIP) told BNamericas,

> Altamira port is very important, it has the capacity to grow enormously, this port has 40,000 hectares of land to be used.[3]

Just as in Mozambique and elsewhere, foreign oilsters and landsters in Altamira will have to negotiate, for turf, with local gangsters.

By 2018 the Citi-controlled land trust also owned hotels in Reynosa (the very center of *increasing Pemex-skimming* by Los Zetas), in Ciudad Juarez, perhaps the quintessential Zeta city in Mexico, and in Veracruz on the southern Gulf coast, controlled by the Cartel del Golfo (Gulf Cartel). In 2012, a year before rollout of the "Energy Reform," the port of Veracruz's entire police force had been dismissed for collusion with Golfo car-

1 HotelNewsNow April 7, 2020

2 FundsSociety Web site, March 4, 2014

3 "Addressing the Concerns of the Oil Industry: Security Challenges in Northeastern Mexico and Government Responses," January 2015 Wilson Center, Mexico InstituteChristian Science Monitor December 22, 2011

telistas. The Mexican Navy took charge of day-to-day police operations in Veracruz.

The appearance, at least, is that Citi sought to do business in towns controlled by the organized-crime cartels, and did not favor one gang over another.

This suspect string of hotel purchases by Citi merely extended a pre-existing pattern of connection to organized crime for Big Real Estate. Before the above hotels were purchased, each of their systems for electronic pre-payment had been breached – by skilled members of organized crime. And so it is remarkable that during the time of these hotel purchases, Citi deliberately was not background-checking employees,[4] including those who handled hotel bookings prepaid by credit card. The U.S. Financial Industry Regulatory Authority fined Citigroup Global Markets Inc. $1.25 million for hiring workers with criminal convictions after Citi failed to conduct background checks on about 10,400 job applicants between 2010 and 2017. By the time of this failure, Citi already had a bad history regarding electronic-payment systems, in that in 2007, soon after Citi took over ATM machines in 7-Eleven stores, that money system was breached by an organized-crime group led by Albert "Soupnazi" Gonzalez.[5]

In the middle of this hotel-buying, Citi sold its "Citi Prepaid Card Services USA"[6] to a then-hot payment company called Wirecard. Wirecard "integrated Citi's system with Wirecard's global payments platform."[7] This integration might sound OK, but, at the time, Wirecard's electronic system was laundering money – on-line casino payments in Malta for 'Ndrangheta.[8]

In character, *Wirecard next tried to buy the bank that administered the Citi-controlled FibraInn trust – Deutsche Bank. Financial Times* called this "a desperate effort to disguise fraud at Wirecard."[9] Wirecard collapsed in June 2020 after being looted.[10] The lame-duck Cheney-Bush administration in 2008 had made a suspect "bailout" of a near-insolvent Citi. The bailout of Citi came November 23, 2008. Reporting was uncritical in the

4 "Citigroup Fined $1.25M For Failing To Flag Criminal Job Applicants," *Financial Advisor* magazine, July 29, 2019,

5 *Los Angeles Times*, August 18, 2009

6 rdc.com Web site, August 27, 2020.

7 Reuters, June 29, 2016, citing Wirecard spokesman.

8 *MaltaToday*, June 13, 2017; Newsbook Web site December 18, 2019; *Financial Times*, August 2, 2020. Currently, Wirecard CEO Markus Braun is jailed on suspicion of money laundering.

9 August 24, 2020

10 *Financial Times*, August 7, 2020 Regarding real-estate mortgages at savings-and-loans, for example, between 1976 and 1990, an alliance among organized crime, spies, bank officials, and real-estate developers looted these institutions. Cf. Steve Pizzo et al, Inside Job: The Looting of America's Savings-and-Loans, McGraw-Hill, 1989

U.S. daily press (citing a "profit" the government could make in a Citi repayment deal).

On March 30, 2010, the monthly and highly regarded student-published *Columbia* (University) *Journalism Review* did some scrutiny and reported that a combined investment in Citi by the U.S. government – in "the bailout" – exceeded the full market value of Citigroup stock on November 21st, the last trading day prior to the deal. In other words, *for the same financial commitment that the government made on that day, it could have owned Citigroup outright.*

U.S. daily reporters failed to investigate this bailout. Such an investigation might have revealed some extent of organized-crime influence on Citi. Reporters also might have examined the havoc Citi banksters were wreaking in Mexico.

Matamoros and Other Gulf Ports

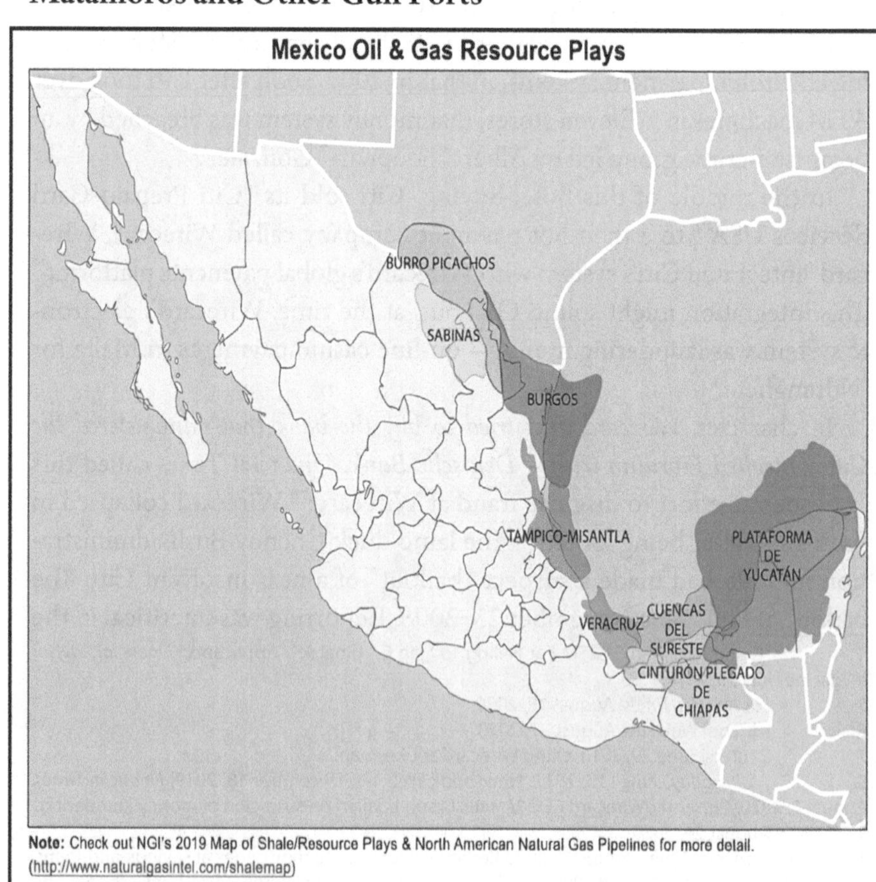

Note: Check out NGI's 2019 Map of Shale/Resource Plays & North American Natural Gas Pipelines for more detail. (http://www.naturalgasintel.com/shalemap)

Source: NGI

The Big Oil and Big Real Estate complex relies on the U.S. intelligence/security community. This was evident in the aftermath of BP's Deepwater Horizon oil spill in the Gulf of Mexico.

Among the Birmingham, Alabama real-estate players earning BP settlement money, BL Harbert Intl. previously had earned U.S. taxpayer money in building a new U.S. Embassy in the Gulf port city of Matamoros. Worldwide in the last decade, BLHI has built more than 30 U.S. embassies – notably in chief drug-trade centers Ciudad Juarez, Tijuana, Honduras, Kosovo, and Chiang Mai, Thailand (the last two are perhaps the world's leading heroin-distribution centers).[11] Routinely around the globe, U.S. embassies house CIA agents covertly or openly, some operatives monitoring and skimming from drug-trade routes.

As such, the following is noteworthy: according to a BLHI staff writer,

> When *BLHI* first arrived in *Matamoros (to build the U.S. Embassy),* its managers were stopped ... by cartel personnel and questioned about their business in the *city*.[12]

I do not doubt that some encounter, its nature unclear, did occur between Los Zetas and BLHI managers on a Matamoros street and that enough people knew about it to warrant BLHI publicizing its own written explanation/interpretation. What is clear is that Zetas in Matamoros knew the faces of the BLHI managers in the Embassy project – either through surveillance or through inside contact. Also influential in Matamoros are the Zeta-allied 'Ndrangheta.

It is likely that in Matamoros, no Embassy would have been built without Mexico's Texas-brokered "energy reform," the privatization of Pemex. In addition to an Embassy, this privatization has brought a major expansion of the Port of Matamoros, which "is expected to improve links with the United States."[13,14]

To the south, at Veracruz in 2017, just as a double-size port expansion was accommodating "a number of large international oil companies," Interpol charged a former Veracruz police chief with *organized-crime* money laundering to finance the chief's buying of five Texas properties and a Veracruz hotel.[15]

11 U.S.Embassy.gov Web site, May 20, 2020
12 EngineeringNewsReview Web site, September 17, 2020
13 Ibid.
14 *Offshore Magazine* Web site, June 13, 2017
15 Associated Press, February 3, 2017.

The misleading label "energy reform" refers to opening publicly owned Mexican oil to private investment capital.

Under the capitalization ("privatization"), President Enrique Pena Nieto granted BP, Halliburton, and Texas's Murphy Oil lucrative contracts – even though in a lawsuit Murphy had admitted buying Pemex condensate found to be stolen by Zetas[16]; as such, likely Halliburton, BP, and Murphy could have done no better if they had written the legislation themselves called "Energy Reform."

And effectively, they had, in that a leading member of their oil-and-real-estate interest group, Citibank/Citigroup, previously had laundered cartel millions in bribes to Mexican officials[17] and in 2013 succeeded in privatizing a Mexican real-estate-investment trust that specialized in hotels (FibraInn).

During the 10-year coverup of BP oil damage to Mexico, nothing reached U.S. ears, either, about cartel bribes to Pena Nieto made in October 2012[18] as Pena Nieto prepared to privatize Pemex. Mexican legislators received part of the bribe in exchange for passing Pena Nieto's 2013-14 "energy reform" bill. U.S. news of the bribe waited until 2019, in reportage on the New York City trial of cartel head Joaquin "El Chapo" Guzman. Pemex CEO Emilio Lozaya has said Pena Nieto in 2013 required Lozaya to deliver bribe money (120 million pesos, $5.6 million) to a total of six legislators.[19]

BP's Deepwater spill helped open Mexico's Gulf coast to profits by the complex of Big Real Estate and Big Oil – while contributing significantly to global warming (through immense release of soot, black carbon, when crews burned off hundreds of thousands of gallons of surface oil).

Also in the Gulf area, in the Bahamas global-warming-caused Hurricane Dorian in 2019 performed a similar opening-up to profiteers in Big Real Estate and Big Oil.

16 BuzzFeed Web site, September 28, 2018. Murphy Oil lawyers admitted the buys but denied the company knew the oil was stolen.

17 Raul Salinas Gortari, for whom Citi laundered $100 million in cocaine money in the mid-1990s. Raul is the brother of Carlos Salinas Gortari, Mexican president at the time.

18 Fox61, January 15, 2019, citing CNN Wire.

19 Mexico News Daily, August 20, 2020

CHAPTER TEN

THE BAHAMAS

Bahamas' Abaco Island filled with refugees from Haiti after a 7.0 earthquake followed four global warming-caused hurricanes in rapid succession.

Eight years later, the global warming-caused Hurricane Dorian obliterated settlements on Abaco. For Big Real Estate, this was the perfect storm, because legally the refugees were squatters on Abaco, and within days, the government relaxed laws specifically to invite land investors to Abaco. Prime Minister Hubert Minnis said,

> We will liberalize the process of investment for the affected areas.[1]

Within months, the Bahamas were teeming with venture capitalists and hedge-fund operators seeking coastal real estate on the cheap. Seemingly impressed, a Florida reporter wrote in September 2019,

> Overall, agents say, investors are attracted by depressed prices, and a possible upside fueled by government disaster funding and new capital from the private sector (including) hedge-fund operators who see future profits in apartment buildings and hotels.[2]

Among those from Big Real Estate to arrive, in style, was South African casino-hotel magnate Sol Kerzner, whose Sun City hotel regularly hosts meetings of African development agencies. South Africa's 2003 hosting of African Union used Sun City as convention venue for delegates with goals of "development," "conflict resolution," and "reconstruction." Resolving conflicts is done to provide stability for investment, and reconstruction is the profitable building of new edifices and infrastructure after conflict. Sun City was part of the resolution of the apartheid conflict. Developer Sol Kerzner built the resort on ground called "homeland" and relegated to settlement by black South Africans. That is, Kerzner at Sun City and elsewhere already effectively had packaged the 2003 African Union goals into his career of simply "making money in the 'developing' world." Kerzner has been called a Steve Wynn/Donald Trump-type character.

1 Relief Web, January 20, 2019; *Los Angeles Times*, January 9, 2017
2 *South Florida Sun-Sentinel*, September 20, 2019

As such, it is fitting that the 2003 African Union conference at Kerzner's Sun City was hosted by one who later would co-author the neocolonialist fake-loan scam in Mozambique – Basetsana Thokoane – and her boss, Foreign Minister Nkosazana Dlamini Zuma (Thokoane would graduate between 2003 and 2010 from press secretary to Defence Department official to intelligence-agency operative; Dlamini Zuma is ex-wife of politician Jacob Zuma).

Kerzner and his lead man, Donald Barrack, Jr. in 2016 moved Kerzner's Atlantis Resorts International Holdings to the Bahamas – just at a time when Barrack was being hired by U.S. presidential candidate Donald Trump, as senior campaign advisor. Trump at one point owned a majority share of the parent company, Paradise Island Hotel and Casino, that owned the Atlantis resort property. In 2010, Barrack bought $70 million of Jared Kushner's debt on 666 Fifth Avenue.[3] Kushner later avoided bankruptcy when Barrack agreed to reduce his obligations after a request by Trump. Barrack's Colony Capital LLC is part owner of a company (Kerzner International Holdings Ltd.) that has built luxury hotels both in Dubai and in the Bahamas.[4]

After Hurricane Dorian, Big Oil as well moved quickly into the Bahamas, represented by a large-scale Halliburton investor, IFM Investors Pty. Ltd.,[5] which bought Bahamas Oil Refining Company International (one of the world's largest petroleum-storage facilities).[6] Then, just five months later, when a separate drilling company, Bahamas Petroleum Company, established a Houston office, it hired Halliburton to frack offshore in the Bahamas.[7]

The Bahamas now is a model of the global-warming future as brought by the complex of top land investors and oilsters.

Shift now back to gas-rich northeastern Africa, target for exploitation by Halliburton in Mozambique and in its neighbor, Tanzania, where Halliburton maintains an office in the capital city, Dar es Salaam.

3 *New York Times*, June 13, 2018

4 Reuters, May 10, 2019

5 IFM Investors owned Halliburton shares worth more than $500,000 as of early 2020. Fintel Web site, January 13, 2020.

6 Along with parent Buckeye Partners Ltd. GlobalNewsWire.com Web site, November 1, 2019

7 Regulatory filing, Bahamas Petroleum Company PLC, January 30, 2020

Chapter Eleven

Tanzania

Tanzanian leader Julius Nyerere with Fidel Castro and a Cuban worker in 1977. Julius Nyerere Archives

According to U.S. media reports, the reason that John Magafuli, recently deceased president of Tanzania, was a source of annoyance to Western officials was that Magafuli was "authoritarian" – a blocker of "press freedom" and "human rights" in Tanzania. These are old media saws, of course, that date back 70 years, to the birth of U.S. anti-communism. America's power establishment, over generations, has never forgiven Tanzania for being friendly with the Cuban revolutionary regime begun in 1953 and won in 1959 by Che Guevara and the Castro brothers, Raul and Fidel.

Tanzanian President John Magafuli was right next door to Mozambique as the MozamScam coopted first Armando Guebuza and then Filipe Nyusi, Magafuli's counterparts in the Mozambican presidency. Half of the gas rich Ruvuma Basin lies in Tanzania, bordering Mozambique's Cabo Delgado where Texas oil, Mozambican officials, and spies were scheming to control the resource.

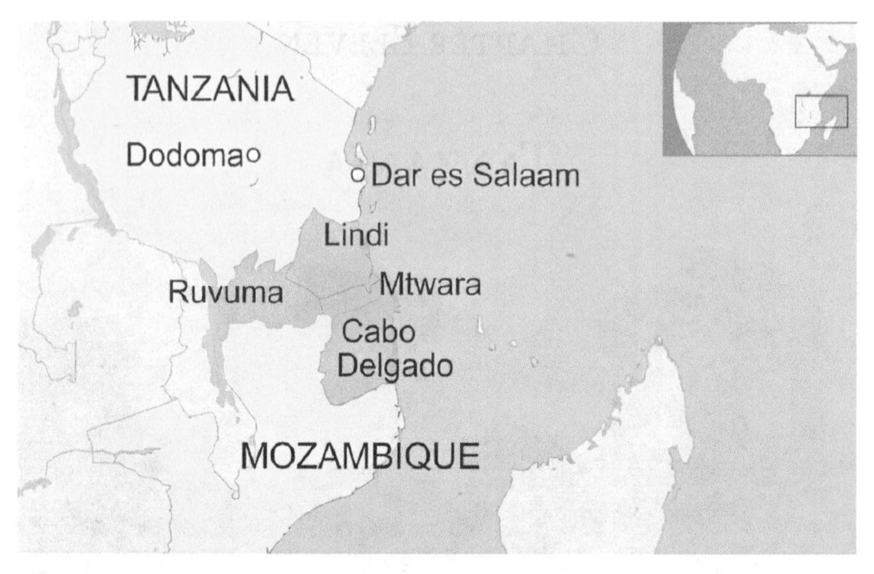

Gas deposits in Mozambique and Tanzania are split by the Ruvuma River (center) on its way to the Indian Ocean

Magafuli was in good position to assess the situation in Cabo Delgado in the same way as do prominent scholars including Dr. Joseph Hanlon, formerly of the London School of Economics,[1] who said,

> (The so-called "Islamic insurgents" in Cabo Delgado) are local people complaining about resource money. Mozambique is a resource-cursed country.... The local elites take the money and don't share.[2]

Magafuli realized that Tanzania is resource-cursed, as well, and drew criticism for not allowing the U.S. to cozy up to him. Magafuli like president Julius Nyerere before him saw that in Cuba, the U.S. had coopted Fulgencio Batista into allowing U.S. interests and Batista's fellow members of Cuba's elite to share profits from developing Cuba's resources while allowing the citizenry to remain in poverty.

This battle with still-unidentified insurgents has been called by experts a "new" type of warfare. Scholar Francisco Almeida dos Santos writes,

> In this type of war, the warring parties *gain from ongoing fighting* and not from victory.

and

1 Hanlon now is affiliated with The Open University, which teaches some classes on the campus of London School of Economics.

2 Interview with Lester Kiewit, CapeTalk Radio, South Africa, April 13, 2021

> "Interests of *private security enterprises associated to investment in extractives* must be considered."[3]

That is, in the instability brought by civil war, security companies profit. Serving resource extractors in Mozambique and Tanzania, dozens of private security companies – unregulated – hire impoverished youths as contractors, then provide substandard food and housing. Also profiting from instability in civil war is "illegal trading" – in extracted rubies, poached timber and ivory, and especially heroin.[4] *Heroin trading requires "mules," carriers.* Heroin dealers hire the same class of youth – impoverished, disaffected – as do security companies.

That is, inhabiting and roaming the Ruvuma Basin are small bands of underfed youths, sporting a rifle here and there, that scare and beat illegal small miners in the interest of corporate mine security. In the same forest, small underfed bands move drugs and scare off both interlopers and police with a rifle or two. And in that same forest, seeing no future in a Western-run economy, more disaffected youths with a few rifles travel toward a mission of war.[5] This is all a matter of public record.

As such, the following is speculation, but I believe it is well-founded.

In Cabo Delgado, young men are both able and motivated economically to rotate among, or to hold simultaneously, contract positions in the unregulated mine-security industry, in the heroin industry, and in the "insurgency." They can do this because they are experienced intimidation-for-hire – contract frighteners. Benefiting simultaneously extraction bosses of all types and heroin bosses, this is all part of a necessary turf rapprochement between heads of industries – some Western, some local, some legal, some illegal – in Cabo Delgado.[6]

"Security" at Tanzania's Chief Diamond Mine

In similarly gas-rich Tanzania, it is reasonable to suppose, and the record suggests, that Western interests long have worked to exploit extractable resources in a way that excludes the citizenry – and even the government. In 2010, Tanzania passed a new mining law, because, as Reuters put it,

> "The mood of the day in Tanzania is that foreign investors *are stealing* (diamonds, gold, and other minerals)."[7]

---

3 Emphasis mine. Francisco Olmeido Dos Santos, CMI Insight, May 2020

4 M. Kaldor, "In Defense of New Wars," London School of Economics, 2013. Also cf. Environmental Investigation Agency Web site February 7, 2013

5 It is likely some ex-Renamo fighters and weapons have joined local youths.

6 Need for rapprochement outlined in Chapter 1, MozamScam.

7 Reuters, April 24, 2010,

In 2017, Tanzania confiscated $15 billion in reportedly undervalued gemstones from foreign-owned Petra Diamonds. In a joint venture with the government, Petra operated and owned a share of Tanzania's Williamson diamond mine. At the time, Petra was in massive debt, with officials ditching company stock,[8] and on or around April 8, 2020,[9] Petra officials closed the premises of the Williamson Diamond Mine.

Three days later, a mysterious "informational" broadcast emanated from the closed Williamson mine premises. It gave false, alarmist Covid statistics. A motive that Petra officials had in the matter was that they likely could have restructured debt if lenders had the impression that Covid, not Petra officials, caused failure to turn a profit.

Police traced the Covid broadcast to the closed Williamson mine via a mobile device being used on the premises by Mariamu Jumanne Sanane. She was arrested under a law barring inflation of Covid statistics.[10]

What was going on here?

Since the mine premises was closed, it is practically certain that Jumanne Sanane had an inside link to these near-bankrupt[11] Petra officials. It is likely this link was forged by Western spies at work routinely in socialist Tanzania; Jumanne Sanane attended a Princeton-sponsored program at University of Dar es Salaam, historically CIA-connected.[12] When Henry Kissinger advocated the CIA recruit students and install professors at foreign universities, University of Dar es Salaam participated, for example during the tenure of sociology professor Dr. Stephen Lucas, a CIA agent, during the 1970s, when Lucas was able to form a relationship with President Julius Nyerere, who was an anti-colonial political activist.[13]

This spy program at the Williamson mine, using Covid to destabilize the rule of President John Magafuli, continued in winter 2021 via "floats" in Western-owned African media[14] and then in U.S. media[15] implying essentially that, by not matching Western Covid practices, Tanzania was "endangering the world." Official cause of death for Magafuli was "cardiovascular disease (CVD)." One researcher in 2011, mentioning "potential association between chronic heavy metal exposure and CVD, said, "Traditional risk factors fail to account for all deaths from CVD." [16] The CIA

8 SimplyWallStreet Web site
9 Mining.com Web site September 9, 2020
10 Xinhua News Agency, April 11, 2020
11 *The Citizen* news@tz.nationmedia.com, July 7, 2020
12 G. Mwakikagile, "Africa in Transition: Witness to Change," *New Africa Press*, 2018..
13 Cf. L. Lupalo, *Nyerere Remembered*, CreateSpace Independent Publishing Platform, 2016.
14 E.g., Quartz Africa, February 16, 2021
15 E.g., *Financial Times*, February 18, 2021
16 E. Alissa, "Heavy Metal Poisoning and Cardiovascular Disease," *Journal of Toxicology*, Sep-

has a long record of assassinations and planned assassinations of foreign leaders, including dozens of attempts to kill Fidel Castro, several of these attempts in partnership with the Mafia. The North Korean government accused the CIA of using nanopoisons in a foiled 2017 plot against ruler Kim Jong-un.[17]

Next in the Williamson mine spy saga, nearly a year after the Jumanne Sanane broadcast, on March 8, 2021, a Kenyan publication suggested Magafuli, incognito, was ill with Covid in a Nairobi hospital."[18] This Kenyan publication, called *The Nation*, is owned by Geneva-born Shah Karim Al Hussaini (the 4th Aga Khan), one of whose enterprises receives USAID funding and another of which, a development bank, specializes in loans to private electricity companies in Africa.[19]

This subterfuge is unsurprising; electricity in Tanzania long has interested U.S. corporations, government officials, and spies. For example, like the Aga Khan's development bank, the Cheney-Bush-founded Millennium Challenge account[20] also receives taxpayer money through USAID that assists African electricity development.

In 2001 – just while Cheney's secret Energy Task Force was mapping the Earth's energy future –Tanzania had a so-called "electricity crisis." This plight was not millions of Tanzanians living without electricity; rather, it was Westerners having insufficient electric power to run modern electronic commerce in Tanzania – commerce done in new Western-style buildings lit, heated, and cooled by electricity.

As USAID put it,

> Electricity demand in (Tanzania) is increasing rapidly mainly due to accelerated productive investments.... Hence, the government is encouraging investments to increase available generation.[21]

The government owns – that is, the Tanzanian people own – the country's electric utility. Not surprisingly, critics said the main "crisis" point was precisely that electricity was publicly controlled. [22] Finnish officials noted in 2001,

---

tember 2011.

17 *Guardian*, May 5, 2017

18 *New York Times* was forced to report the existence of this "rumor." Historically, a reputable publication will avoid mentioning rumor until substantiated.

19 Aga Khan Foundation and Aga Khan Fund for Economic Development, respectively. Wikipedia

20 Approved by Congress in 2004.

21 USAID, "Investment Brief for the Electricity Sector in Tanzania"

22 *Journal Energy Southern Africa*, May 2018. Later, in Mexico, a privatization scheme opened state-owned Pemex oil company to private investment; see Chapter xx of this book.

> *There are plans to privatize the utility company (TANESCO) but it may be difficult without further structural changes.* The Government is in principle prepared to take the necessary steps to *correct the situation* but there may not be enough *political strength* to do so.[23]

Opposition to "structural changes," and lack of "political strength" appear to be code phrases for a perceived lack of wholehearted enthusiasm for Westernization and corporate capitalism held by Tanzania's socialist government.

At this time, 2001, this kind of "problem" likely was just the *kind with which Cheney's Task Force concerned itself.*[24] The overall plan, the record indicates, included *pushing Tanzania away* from climate-friendly hydropower to global-warming production of electricity by burning the country's natural gas – developed by U.S. interests.

Right away, in March 2002, the Cheney-Bush administration proposed its new tool for global development, the "Millennium Challenge Account," packed with $3 billion per year in U.S. taxpayer money.[25]

The MC account is used, among other things, to weaken socialism abroad. As the U.S. Council on Foreign Relations put it,

> MC funds will go to countries that enact market-oriented measures designed to open economies to competition...

This opening to capitalism clearly included the 110 trillion cubic feet of natural gas deposits in the Ruvuma Basin joining Tanzania and Mozambique.[26] The first step was for Texans to intervene in Tanzania's plan for hydropower, with an eye to derailing the hydro plan in favor of U.S.-developed natural gas.

In 2006, Houston entrepreneur Mohamed Gire, proprietor of Richmond Printing, won a contract to supply Tanzania with hydropower. Gire managed this with nothing but a shell company[27] and political influence – after Gire got the hydro contract, a Tanzanian government minister's company bought the contract. After that, the contracted hydro turbines failed to produce electricity,[28] but the government minister's company got the government to pay the contracted money anyway, $4 million a month for two years (some of this likely kicked back

23 Ministry for Foreign Affairs of Finland, 2001, emphasis mine.

24 Later, in Mexico, the same kind of problem-solving by Texas oilmen would succeed in privatizingexico's publicly owned oil utility, Pemex; see Chapter xx of this book.

25 U.S. Government Printing Office, Senate Hearing 109-187

26 Mozambique got its first Millennium Challenge money in 2007, Tanzania its in 2008.

27 An on-paper-only offshoot of Gire's Houston-based Richmond Printing Company

28 U.S. Embassy Cable 08DARESSALAAM98_a, February 8, 2008, published by WikiLeaks

quid pro quo to Houstonian Gire's Richmond Printing Company and/or a slush fund behind him).

U.S. Tanzania Ambassador Mark Green reported,"[29]:

> "Pressure (was brought) on the Tanzanian Electric Supply Company (TANESCO) to circumvent the required public procurement process and sign a contract with the Houston-based Richmond Company for temporary energy relief during Tanzania's 2006 energy crisis. Richmond had promised to deliver 100 megawatts of power within two months; not one watt of energy was ever produced.... Since Richmond is a U.S.-based company, the Embassy received many inquiries, but could only say that Richmond was registered in Houston Texas as a "printing shop and business services center."[30]

This was disingenuous. The record shows Gire at the time had friends at the highest levels of power in the U.S.. He was making more such friends every day. Gire in 2010 was named to Barack Obama's Presidential Summit on Entrepreneurship. At that time, Gire's business associate[31] Lutfi Hassan already had been Obama's national finance co-chair for the Obama for America campaign of 2008.[32]

Just at this time, of course, petroleum interests were trying hard to shift Tanzania from hydropower to natural gas. A whopping $1.2 billion award from the Cheney-Bush Millennium Challenge account[33] made the U.S. a controlling "partner" in Tanzania's power industry between 2008 and 2016.

It is worth asking whether the fraud by Houston-based Richmond was connected to a firm surmise that Tanzania's water supply would prove inadequate for hydro turbines to power the country. Such a surmise easily could have been made as early as 2001 by the Cheney Task Force members – as we have seen, Cheney Force members were experts, of a type, on global warming – some of the best-informed in the world on the extent and causes of warming. At the time, extended drought was somewhat predictable in Tanzania: "Atmospheric circulation patterns observed in the Tanzanian region from 1998 to 2005 were similar to those experienced during a previous prolonged drought (1973 to 1976), suggesting... predictability of drought."[34]

29 Ibid

30 Ibid.

31 Through Apex Worldwide, which Hassan ran and for which Gire worked; ApexGroupCompanies Web site.

32 Ibid.

33 Center for Strategic & International Studies Web site, May 2, 2016

34 Agnes Kilazi, "Analysis of the 1998 to 2005 drought over the northeastern highlands of

Sure enough, Tanzania conceded in 2014 that global warming had made hydropower unreliable in Tanzania and the nation would switch to gas power. A "reform roadmap" document[35] said,

> Many of TANESCO's problems were caused by the dire effect of climate change on East Africa's previously predictable rainy seasons, curtailing output of hydropower plants....(Therefore) the Roadmap...envisages large increases in gas and coal.[36]

Thus, although TANESCO was still public, Tanzania had made major concessions to the extractors of gas and coal who, working with Millennium Challenge money, nearly had gotten TANESCO privatized. This of course recalls the successful privatization of Mexico's Pemex. Following the thesis of deliberate global warming to profit investors, this concession was a victory for such investors. Investors were set to make money while Tanzania committed to accelerate burning of its 57 trillion cubic feet of natural gas, thereby accelerating global warming.[37]

John Magufuli

For the record, Erik Prince's FSG works in Tanzania,[38] and Halliburton operates an office in Dar es Salaam.

As of mid-2021, it was clear that Magafuli's death was a boon to Western extractive interests and to investors in those companies. Magafuli resisted Westernization. He may have been murdered.

His successor, Samia Suluhu Hassan, is proving markedly less resistant.

Westerners Challenge Tanzania's Constitution

Notably, Western-owned and -supported media has urged the Suluhu Hassan government to abandon Tanzania's current Constitution. This is, of course, a radical demand, and it comes from the same Western-supported press that carried the false story of Magafuli's hospitalization in South Africa with Covid.

"Post-Magafuli era" is code for the agenda of Tanzania's political opposition, Chadema, defined in Wikipedia as "a radical center-right party." Chadema calls itself a "free-market party." Tanzania has long had just a

Tanzania," *Climate Research*, March 2009

35 Tanzania's "Electricity Supply Industry Reform Strategy and Roadmap."

36 In part through Feed the Future Innovation Lab for Food Security Policy (FSP) Agricultural Sector Policy and Institutional Reform Strengthening (ASPIRES) project.

37 I.e., this of course recalls the successful privatization of Mexico's Pemex.

38 AfricaIntelligence.com Web site, July 24,2020

single party, committed to maintaining autonomy from colonialism. At bottom, the Chadema faction, then, is a neocolonialist group. Chadema is led by Tundu Lissu.

Tundu Lissu

On August 18, 2017, Lissu from Belgium through Kenyan television vociferously publicized to Africans that an airliner bought by Tanzania had been seized by a Canadian court in a money dispute with a Canadian road contractor who had worked in Tanzania. Lissu, who is the chief lawyer of Chadema, told the public that if the dispute wasn't sorted immediately, it was possible the government would go bankrupt. He said that all this information was obtained from Chadema insiders in the government, however, he did not name his whistle blowers.[39]

This was a quite effective publicity stunt, gaining worldwide coverage.

Lissu had been charged months earlier with sedition (although charges were dropped). In this context, the Tanzanian government suspected Lissu of having a hand in getting the plane seized. A government spokesperson said it is one of several "dirty games" that involve "cooperating with to sabotage the implementation of several development projects"[40] initiated by President Magufuli.

Three weeks after the publicity stunt, on September 7, 2017, a press release by Chadema, Lissu's opposition group, stated Lissu had been wounded in a shooting attack.

39 The Citizen Web site, August 19, 2017
40 Lexology.com Web site August 31, 2017

Barely two months before this allegation, Chadema had alleged that another opposition leader, Freeman Mbowe, had been attacked and injured. Police investigated, found no evidence of such an attack, and said they "had doubts" Chadema was telling the truth.[41]

Lack of Evidence

Chadema stated Lissu had been shot several times in a hail of bullets in the city of Dodoma and that after a local hospital stay he had been removed to a hospital in Kenya. Notably, the group stated this was done not for the purpose of better medical care for Lissu but rather "out of fear for his safety." News stories reported the shooting as fact; the number of bullets reported varied wildly, from 16 to 60.

Unsurprisingly, Chadema said the government was responsible. However, if shooters did attacke Lissu, they could have been agents of the "foreigners" in whose "dirty games" against Magufuli Lissu allegedly was playing – Lissu's publicity stunt around the seized aircraft could have been viewed as an out-of-control action by a man who knew too much.

This is of course speculation, but where evidence is lacking speculation is sometimes warranted. And here, remarkably, despite many dozens of Internet reports issuing, no conclusive evidence was ever mentioned (that I could find) showing Lissu had been shot – only Chadema allegations.

For example, an Al Jazeera Web link no longer works for whatever reason that presumably reported in some depth on what hard evidence there was or wasn't about the alleged shooting of Lissu.

And, the situatiion around the alleged shooting displays numerous remarkable coincidences, so many that together they further call into question the Chadema account.

According to the account, Lissu was taken to the Aga Khan Hospital in Nairobi. This hospital is owned by the Aga Khan Foundation.

Coincidentally, the Aga Khan Foundation runs a development project concerned with Tanzanian electricity, it coincidentally has long been funded by the State Department's USAID, and it coincidentally in 2019 graduated to a full "partnership" with USAID in a cross-border development project in Tajikstan and the Kyrgyz Republic, called "Local Impact" (which set of programs USAID coincidentally plans to expand to African nations).[42] USAID has been called a CIA front.

41 Agence France Press June 12, 2020, The Web link "READ MORE: Tanzania: Opposition MP Tundu Lissu wounded by gunmen"

42 USAID Web site

Lissu was sent from Aga Khan Hospital to the Leuven University Hospital in Gasthuisberg, Belgium. Coincidentally, Aga Khan Hospitals and Leuven University Hospital are affiliated – through collaboration on at least one research project[43] and through their standing as two of only five total "stakeholders" in the International Society for Congenital Adult Heart Disease.[44]

Remarkably, Leuven Gasthuisberg is not a standard surgical hospital. It is a "teaching hospital." At a teaching hospital, patients are treated by resident interns, who hold MD degrees but have no experience treating patients.

What happened there?

Coincidentally, for that account we have only a report in Nairobi's *Daily Nation Kenya*, which newspaper is owned by, well, the Aga Khan Foundation.[45]

Although Chadema originally stated Lissu was moved from a Tanzanian hospitial not for further medical care but merely out of fear for his safety, the *Daily Nation Kenya* reported that Lissu in Belgium received 19 separate surgeries for wounds[46] – these surgeries supposedly done by Leuven teaching-hospital's intexperienced resident interns. If the medical residents at Leuven Gasthuisberg did put scalpels to Lissu numerous times, any bullet holes he had – or the absence of bullet holes – likely are no longer in evidence.

Incidentally, the Aga Khan's *Daily Nation Kenya* was the publication behind the rumor that Magufuli was ill in Kenya with Covid-19, reporting spuriously on the Internet March 10, 2021, "African leader admitted to Nairobi Hospital with Covid-19,"and in print the same day, "African leader taken ill, admitted to city hospital." The stories said the unnamed "leader" was incognito, but they contained paragraph after paragraph about Magufuli's controversial response to Covid. Over Kenyan television from Belgium, Lissu himself said Magufuli died from Covid.

On July 27, 2020, Lissu returned to Tanzania to contest against Magufuli in the general election. The U.S. and other Western nations balked for months at acknowledging Lissu lost in a fair election. Lissu returned to Belgium in "exile."

But almost immediately, *Chadema* and the ACT-Wazalendo political group began talks to unite behind Lissu.

43 i.e., the *Multicenter International* Survey on Cardiopulmonary Bypass Perfusion Practices in Adult Cardiac Surgery , 2019

44 Approach IS-II Web site

45 Wikipedia

46 "Lissu to undergo 19th operation in Belgium," *Daily Nation Kenya*, May 15, 2018

The day after Magufuli's death was announced, a representative of the U.S. Council on Foreign Relations think tank's Africa Program obtained a meeting and interview with the still-exiled Lissu, reporting in a rather odd published piece that emerged from the meeting,

> Following the president's untimely demise, Lissu is plotting his return.[47]

It seems the CFR think tank, once directed by Dick Cheney, might have originally intended the meeting with Lissu to be a secret briefing by Lissu to American higher-ups on what his plans were, and how subversive they were of the Tanzanian government.

As it came out, the interview was deemed fit, even advantageous, for publication, and the piece opened with the following quote from Lissu:

> "Well, we are not trying to overthrow the government, are we? No? Right, so let's do it on the record."

This reminds one of dialogue between spies in a John LeCarre story who believe they are being listened in on.

Then, CFR rep Nolan Quinn proceeded to write,

> Tundu Lissu, a Tanzanian opposition leader, does not aim to overthrow the government, but he nonetheless has big goals for his home country. Following the death of Tanzania's authoritarian President John Magufuli, an avowed COVID-19 skeptic who likely died of the disease, Lissu plans to return to Tanzania to fight for democratic progress in a country that has experienced rapid democratic backsliding in the last five years.

Lissu's and Quinn's assertion of "no aim to overthrow" is debatable. The story confirmed that Lissu wants Tanzania to abandon its current Constitution in order to do away with "the imperial presidency put into place by Julius Kambarage Nyerere."

While president from 1964 to 1985, Nyerere argued that because African nations were in a post-colonialist stage, political parties in Africa were different from parties in the West, saying,

> The European and American parties came into being as the result of existing social and economic divisions - the second party being formed to challenge the monopoly of political power by some aristocratic or capitalist group. Our own parties had a very different

47 A CFR Web site

> origin. They were not formed to challenge any ruling group of our own people; they were formed to challenge the foreigners who ruled over us. They were not, therefore, political "parties," i.e., factions, but nationalist movements. And from the outset they represented the interests and aspirations of the whole nation.[48]

Today, it is clear that the Chadema "party" represents a neo-colonialist faction in Tanzania.

New President Samia Suluhu Hassan has not agreed to abandon the one-party Constitution. Instead, she has conciliated Chadema by inviting in neocolonialist projects, saying,

> Let me improve the economy, and other things will follow.[49]

Samia Suluhu Hassan

Suluhu Hassan has resumed talks aimed at bringing massive foreign investment in developing Tanzania's coastal city of Bagamoyo into a mega-seaport that will significantly accelerate global warming. Magafuli had delayed this project, which will include an essentially private "New City." Western media uniformly says prospective investors are "Chinese." This is almost entirely inaccurate.

Controlling the new city and its 800 hectares of Tanzanian coastland will be the Omani Investment Authority, OIA, which caters to foreign investors (Oman is a sovereign state in the United Arab Emirates)[50] and China Merchants Port Holdings, a Hong-Kong-run company, in which private investors own a large majority of shares. Only around a third of shares in CMPH are publicly owned by the People's Republic of China.[51] As such, it is inaccurate, although still well-believed, that CMPH is "a Chinese company."

Hong Kong, like Dubai, is a notorious "soft spot" in capitalism where projects often include both "legitimate" and "illegitimate" capital. Omani Investment Authority is a strategic-development sovereign-wealth fund that among other ventures offers foreign investors shares in numerous coastal-tourism resorts.

So, through CMPH and the OIA, any investor, including Americans, could participate in controlling Tanzania's Bagamoyo port. He or she

48 Quoted by J. Quigley, "Perestroika African Style: One-Party Government and Human Rights in Tanzania," Michigan Journal of International Law, 1992

49 Deutsche Welle, June 29, 2021, "Opinion: Samia must bring Tanzania to post-Magafuli Era," by Sylvia Mwehozi

50 SWF Institute Web site, January 31, 2021

51 Bloomberg, April 28, 2020

could even quickly show that in spite of mainstream press, Bagamoyo, rather than being "a Chinese project," is a Western-style project wide-open to U.S. and other investors.

In fact, CMPH and OIA have imposed *an operating condition saying that Tanzania should not question whoever comes to invest there once the port is operational.*[52] When Magafuli in 2016 delayed the project, he had in mind this onerous condition. Clearly, this loophole could accommodate organized-crime investments. A further operating condition states Tanzania must not build any newer port after Bagamoyo, on a large majority of its coastline. Hence, investors recruited world-wide to the Omani Investment Authority/China Merchants Port Holdings investment fund will have a kind of veto control over a majority of Tanzanian coastline real estate.

One further hitch was that Tanzania under Magafuli cared about coastal landowners whom the Bagamoyo port project will displace. The government had demanded investors pay displaced landowners a total of $28 million as part of the cost of doing business in Tanzania. Instead, in exchange for $28 million, investor CMPH demanded and got Tanzania's entire share of ownership in the Bagamoyo project. Investors squeezed Tanzania out. As it was reported in July 2018,

> The China Merchants Holdings International Company has agreed to compensate the landowners, which has cost the Government its share of ownership in the (Bagamoyo) project.[53]

All of this means Tanzania's coast under new president Suluhu Hassana will be forced to endure a Bagamoyo "Investor City" parallel to the "Gas City" being built in neighboring Mozambique.

Tanzanian Gas and Exxon

Exxon has secretly planned[54] *that by 2025, its carbon dioxide emissions will rise sharply – by a full 17 percent* (which is more than the current emissions of the entire country of Greece). In this context, Exxon is set to drill for Tanzanian gas. Exxon holds a 35 percent stake in Tanzania's deepwater Block 2 field, which holds an estimated 23 trillion cubic feet of gas. Unlike Exxon, BP and most other oil companies have pledged to cut carbon emissions. This exposes Exxon as a rogue, an outlier, in

52 GhanaWeb, June 27, 2021
53 Zawya.com Web site July 31, 2018
54 *USA Today*, October 5, 2020, according to leaked documents first reported on by Bloomberg.

the current scheme of things. Previously, Texas oil companies had seen BP as a rogue when BP admitted oil companies caused significant global-warming (as discussed in Chapter 6, a stance by BP that may have caused Halliburton and Transocean to set up BP for the Deepwater Horizon blowout).

This revelation concerning Exxon's duplicity, combined with recent evidence that oil companies' estimates are far too low on emissions in the first place, is the kind of dire matter that regulators and legislators deserve to be notified of, voluminously, in messages from citizens.

Digression: Carbon-Trapping Technology

It is worthwhile here to digress briefly. EPA regulators also deserve heads-up that Exxon plans, as well, to adopt carbon trapping technology.[55] This move will not reduce Exxon's planned 17-percent raise in carbon emissions (so the move is at best scarcely laudable) but the concept is sound. This trapping technique binds carbon molecules, preventing escape to the atmosphere, much as forests do in photosynthesis. Carbon-trapping will take money, but the technology works, and experts today are saying that carbon-trapping is necessary to win the climate struggle.[56] So, the flood of messages deserved by regulators at the EPA arguably should mention the obvious, that EPA needs to mandate significant spending by oil companies on carbon-trapping technology.

Big Real Estate and Tanzania

As we have seen in the case of Mozambique, oil exploration tends to usher in other developers, including those who will build roads and hotels in Cabo Delgado's scheduled "Gas City." Concerning the Tanzanian gas deposits pictured above (lying just north of Cabo Delgado), none of them, from Mtwara in the south to Kilwa Masoko on the central Tanzanian coast, can be drilled without installation of massive infrastructure – roads, pipelines, corporate buildings, parking lots, hotel accommodations, a 90-mile-long, global-warming Gas City for Tanzania – which will require a depopulation of the rural coastline. This invites in Big Real Estate.

Alongside duplicitous oilsters, landsters figure large in the future of Tanzania. The Tanzania Investment Act of 1997 invites foreigners to purchase land, solely for investment purposes.[57] Between the cities of Dar es

55 Ibid. to 35.

56 Announcement by the International Energy Agency, reported in the *Guardian*, September 24,2020.

57 For personal use of land, foreigners must purchase through a Tanzanian resident.

Salaam to the north and Lindi to the south, most of Tanzanian coast hosts small villages, of fishers and farmers. Already investors are buying up land just uphill from coastal Dar es Salaam.[58]

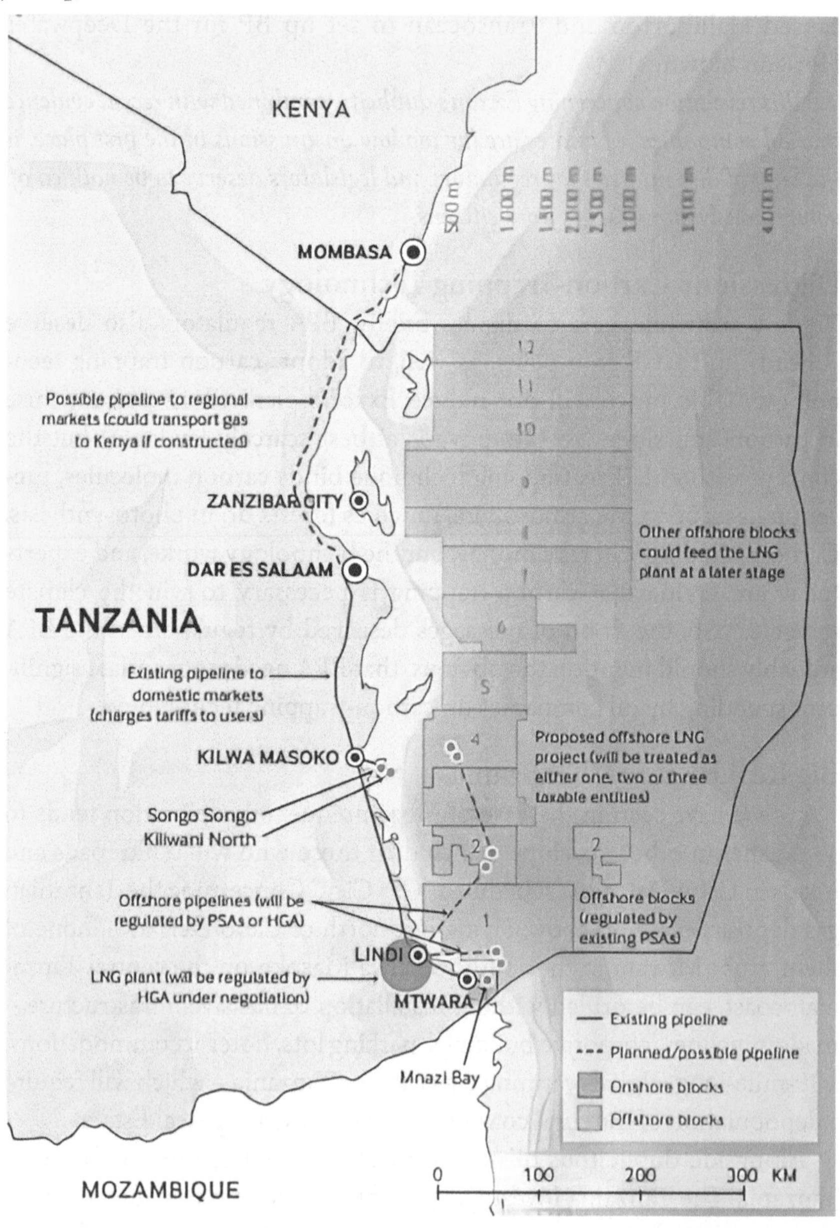

It is likely that some capital invested by foreigners in real estate on the Tanzanian coast is ill-gotten gains (as on the Mexican Gulf coast).

58 Gb Marketing and Real Estate Co. Web site

U.S. hypes ReMax

AllAfrica Web site, July 16, 2021: The United States Wednesday commended Tanzania for the efforts the latter is making towards attracting more foreign investment in the country.

U.S. ambassador to Tanzania Donald Wright made the remarks as he reacted to recent measures by Zanzibar to initiate a programme that would allow foreign investors to more easily take advantage of tax and residency benefits for investment on the Isles.

The ambassador explicitly plugged ReMax. ReMax is global Big Investment Real Estate writ large. A 2017 class-action suit alleges ReMax issued materially misleading business information to the investing public.[59]

"With Re/Max's expansion in Tanzania, U.S. and other investors will have a globally recognised advisor to assist them in making these important business investment decisions," noted Mr Wright.

Explaining about the incentives, Zanzibar's Finance Minister Jamal Kassim said the revolutionary government had recently passed the attractive policy incentive to the real estate industry.

He said for investors coming with an investment capital of at least $100 million (about Sh230 billion), they were now enjoying exemption on import duty and Value Added Tax (VAT) on equipment meant for the project.

The public message from the U.S. ambassador seemed aimed at hyping a coming U.S.-Africa Business Summit presented by the Corporate Council on Africa, an organization founded by the State Department's USAID.

Those who hold real-estate investment-trust stock could do the same as could investors in oil stocks, in this case notifying regulators about unreliable estimates of Sea Level Rise that are used illegitimately to gain investors. The Securities Exchange Commission regulates domestic investment trusts, and the U.S. Foreign Asset Tax-Collection Act (FATCA), which governs foreign Real Estate Investment Trusts (REITs), is enforced by the SEC and the IRS.

Increased migration of rural folk itself has already accelerated global warming as cities have expanded. Around this has arisen an anti-immigrant narrative, fueled by Donald Trump, which narrative has increased social anxiety. A more-anxious population is readier to buy real estate – either for a comforting "sure" investment or for the comforts of a new house.

And let us not forget the hard influence of organized crime. For this, Tanzania is an introduction to affairs around the African continent.

59 Rosen Law Firm press release, November 3, 2017

Chapter Twelve

Around Africa: Little But Westernization, Little Else in Store

Maintaining indigenous culture

Inseparable-Triplets, Trends for Africa: Westernization, Organized Crime, and Global Warming

Westernization of Africa is being accomplished in general by construction of what are called "New Cities." These are cities built largely by development corporations whose primary ownership is Western. More than three quarters of New Cities across Africa are 100-percent privately funded. Combined investment in African New Cities currently is around $115 billion.

A concern with these Western-style New Cities is that, from concept to construction, they silence or even outright exclude local urban planners and managers, local governments, and the citizenry itself. Many New Cities are being built with input exclusively from architects, engineers and property developers."

As executive Ronald Chagoury of South Energyx Nigeria told the *Financial Times*,

> Multinationals that want to penetrate the Nigerian market must have a local head office, and our city offers a good location for them.[1]

The Rendeavor company – whose two largest investors are non-Africans and are billionaires – is developing a total of seven new cities across the African continent. These are Alaro City and Jigna in Nigeria, Apollonia City and King City in Ghana, and one new city each in Kenya, Zambia, and Democratic Republic of the Congo.

Together, these New Cities will cover nearly 50 square miles, much of it deforested.

1 FDI Web site, December 27, 2020

Internet

It is likely that most residents in the New Cities will be Internet users. The same goes for new residents of Dar es Salaam, Tanzania, the world's second fastest growing city. Dar has six million people but is expected to have 10 million fairly soon. Internet "penetration" already stands near 50 percent in Tanzania, with a great recent rise in number of smart-phone users. Most of these, it seems, enjoy on-line gaming, which has already made a mark in Tanzania's revenue picture, described as follows:

> A recently *recorded massive growth in tax collections from gaming activities, due to the rising number of players and tighter controls.*[2]

A notable share of this gaming-tax revenue comes from on-line casino gambling, which jumped in Tanzania when sports betting was legalized in the late 1990s.[3] Casino games are still not completely accepted everywhere in Africa, but they are proven popular in Tanzania, South Africa, Kenya, and Nigeria.

"The first one's free."

A particular on-line casino company, Gambino Slots, believes it can tempt Africans into frequent on-line gambling simply by offering games where players don't play for money. A free 200-coin "stake" is issued by the company.

The company says,[4]

> In terms of being free, there is no better temptation than that.

Once the free stake is used up, however, players must buy "G-coins" for the slots, to continue gaming just for thrills.

The company says,

> In essence African players, both mobile and desktop users, are benefiting from the success of brands like Gambino. Gambino saw the trend and has integrated their wide database of games into 'app' production, releasing all their games to the mobile public. Gambino Slots brings a Las Vegas feeling online, but it is not really gambling. The main purpose of the games is to have fun and try your luck in a risk-free environment of your own home.

2 James Mbalwe, Director General of the Gaming Board of Tanzania, stated in 2019

3 The Citizen Web site

4 Gambino Slots Web site

The company avoids mentioning potential habituation to gambling that use of its games entails. Technically and legally, a disclaimer is offered by Gambino Slots, but a reader must push an extra on-line button to get to it.

> Practice or success at Gambino Slots does not imply future success at "real money."

University researchers found the following"

> When online gambling becomes a routine daily behavior, it is easier for consumers to engage in 'mindless' consumption of that activity, ultimately resulting in addiction and resultant financial losses.[5]

Study results show online gamblers gamble more frequently and aggressively.

Internet-Casino Gambling and Organized Crime

It is highly likely that choice of the name "Gambino" is meant to connote old-style casino gambling as it was overseen by organized-crime patriarchs, such as New York City's Carlo Gambino. Organized crime loves to profit off the practice of gambling, and Internet gambling is no different.

According to the TechWire Web site, July 9, 2022, a "gaming computer" consumes six times more electricity than a laptop. The importance here of that is that very few online gamblers are aware of it. This fact caused issuance of a warning from researchers at Cal Berkely's Lawrence Laboratory.[6]

5 "New Study Shows Online Gambling More Addictive than Casino ... – University of Nevada, Las Vegas, July 16, 2008

6 These researchers, in "Taming the Energy Use of Gaming Computers," Energy Efficiency, 2015, by Lawrence lab scientist Evans Mills et al, found the following:

A gambling computer equals three refrigerators' power use, making it the major appliance in a gaming household, and

Annually worldwide, gaming PCs drain the output equivalent of 25 typical power plants, and by estimate this was to double by the year 2020.

Online gambling gives a *criminal* actor the potential to be even more anonymous than in physical *casinos*. Casino Web sites are often based in tax shelters that have favorable privacy laws. Organized-crime groups have bought up online-betting venues and have founded their own gambling sites.

In Malta, owners of a vast gambling network tied to La Cosa Nostra were arrested in Sicily in 2018; soon after, Calabrian Police seized a Maltese betting company because of its links to 'Ndrangheta. [7] The reader will recall the connection of 'Ndrangheta, Malta, and online casinos from the Chapter Six discussion of how the Skanska firm won US contracts despite being implicated by a Maltese judge in a scheme with organized crime in Malta.

Smart-Phones and "Conflict Minerals"

Many on-line gamblers use smart-phones. In Tanzania alone, more than 30 million smart-phone users are expected by 2024. The manufacture of 30 million smart- phones consumes some substantial weight of minerals such as cobalt, tungsten, and tantalum. All three of these minerals are found and mined heavily in African nations. That is, for each new Tanzanian gambler-by-smart-phone, tantalum will need to be mined in Rwanda or the Democratic Republic of the Congo.

Tantalite, the ore bearing tantalum, is traded on a market that is one of the world's most secretive. It is not traded on commodities exchanges. Instead, it moves through "shadowy networks of dealers" who are adept at concealing its origins. That is, end users – makers of everything from iPads to airplanes – can't tell whether a war zone was the origin of their key ingredient.

Although not in every case, a civil war in Africa frequently is caused chiefly by dispute over which armed force – government or rebel – is going to exploit resources and profit that way. Areas around mines are battlefields with the winner of the most recent battle taking control of mine

---

The point here is that "gaming PCs" are equivalent to bitcoin-mining computers in being stick-out-like-a-sore-thumb contributor to global warming.

Additionally, researchers found evidence that the idea of conserving energy against global warming seems antithetical to the gambler's mindset. Responses to the Lawrence Laboratory article included the following:

Interesting read. Still don't give a shit though because I like to harness thy gigahertz.

and

People above me already said it all; we have capitalism and I can waste my money on dank pc gaming framerates if I want to.

and

I overclock the shit out of my computer most the time which increases power draw to about double what it is at stock. This nets an additional 20 percent performance (average). And then I suffer because it turns my room into a toaster. But I do it for the love of the PC MASTER RACE.

7 OCCRP Web site, March 4, 2021

profits. This functions to extend wars. *It's profitable over the long run for each side to keep fighting.*

It is in this way that tantalum, tungsten, and cobalt have come to be dubbed "conflict minerals." Unfortunately, all of the above means that every increase in cell-phone use – by making mining conflict-mineral mining more profitable – tends to create new and prolong old conflicts in African nations. *As we shall see below, armed conflict accelerates global warming in ways that aren't even acknowledged widely yet. this means every increase in cell-phone use in the West and every increase in Africa will accelerate global warming.*[8] *So far, wide concern has been raised over conflict minerals in the context of human rights but almost none in the context of global warming.*

In 2013, Rwanda dramatically boosted tantalum exports to become the world's largest exporter, shipping out to foreign smelters some 2,460 tons, 28 percent of the global total of around 8,800 tons. That is more than double 2012's exports, despite the fact that Rwanda has consistently produced around 1,500 tons annually, according to United States Geological Survey data.

The spike deepened suspicions that ore was being smuggled across the border from conflict areas in the DRC across the Rwandan border from conflict areas in the Democratic Republic of the Congo. In April 2013, more than two years after the CFSI began auditing smelters, Global Witness reported that

> Much of the tin, tantalum and tungsten produced in the DRC benefits rebels and members of the state army.

This includes the recently integrated rebel group National Congress for the Defence of the People (CNDP).[9]

The US State Department has been influenced by a report titled, "The Islamic State in Congo," which claims that Congo's Allied Democratic Forces (ADF) was connected to the Islamic State and had received war materiel from the group.[10] Scholars who have studied Congo's violence do not support the claims in the report.

A co-author on the report works for Texas-based Bridgeway Foundation, affiliated with the investment fund Bridgeway Capital Management.

8 As acknowledged in the Introduction to this book, Asia – China in particular – will also contribute to this phenomenon. The case of China will be examined later in this chapter, in the context of politico-economic conflict between the US and China, much ballyhooed in the US.

9 "Organized Crime and Instability in Central Africa," United Nations Office on Drugs and Crime, 2011

10 H. Epstein, "The Bewildering Search for the Islamic State in Congo," AfriDesk.org Web site, April 26, 2021

Bridgeway's head, committed Christian John Montgomery, devotes part of his investment company Bridgeway's profits to hunting down and forcing surrender on members of the Allied Democratic Forces. Bridgeway operatives have paramilitary training, hire private intelligence agents for operations, and align their activities closely to US foreign policy objectives.

Within weeks of the report, Secretary of State Antony Blinken designated the ADF – calling it "ISIS-DRC" – as a "Foreign Terrorist Organization." The label opens up Pentagon funding lines.

The ongoing wars in Rwanda and the neighboring DRC has been called the bloodiest since World War II, but armed conflict characterizes numerous African nations; the US Conflict Minerals Law applies to materials originating (or claimed to originate) not only from Rwanda and the DRC but from the eight adjoining countries: Angola, Burundi, Central African Republic, Republic of Congo, South Sudan, Zimbabwe, Uganda, and Zambia. Illicit trafficking in precious minerals has been characterized by a former South African police executive as a "fleecing of national assets and resources." This is disingenuous, because as such illicit trafficking s identical in general to Western corporate extraction of African resources – and all of it speeds global warming.

In wars fought around control of mineral mining, among the antagonists there is a common financial incentive to maintain a credible threat of instability in a region. The DRC war like the Mozambican war is fought not toward victory but toward profit, the fight itself is the victory.

In eastern DRC, illegal resource exploitation generates $1.25 billion per year. Of this, estimates say armed groups take $13 million,[11] representing yearly subsistence cost for at least 8,000 armed fighters – enabling defeated or disarmed groups to continuously resurface.

This means electronics prolongs this and similar wars, which, as we shall see, prolongs accelerated global warming.

11 UNEP-MONUSCO-OSESG Final Report, 2015.

Chapter Thirteen

Militarization Of Africa

In ways that have not yet been widely publicized, armed conflict accelerates global warming. This will be reviewed in the final chapter of this book.

The African continent is being militarized by foreign nations including the U.S. The record shows that if this militarization was intended to keep peace – which is quite dubious – then that intention was misguided.

The U.S. African Command: AFRICOM

Many conflicts in Africa have arisen around the extraction of resources. The U.S. and France, chief extractors of African resources, have claimed this type of conflict as pretext for deploying troops in Africa. Historically, the U.S. has called this type of pretext "protecting American citizens (working in a conflict country)," as it did in the case of U.S. occupation of Nicaragua from 1912 to 1933. Recently in Africa, the pretext is called "training the military" of a conflict country. But the U.S. has recently spelled out the policy underlying the pretext: policy is to deploy U.S. troops to "ensure a steady flow of resources" from a foreign country, or as Africom commander Gen. Stephen Townsend put it"Africa: Securing U.S. Interests, Preserving Strategic Options."[1] ,

> Our future security, prosperity, and ability to project power globally rest on free, open, and secure access in Africa.

This bundling together with "security" and "ability to project power" obscures the objective for Africom of ensuring U.S. corporate profits – "our prosperity."

Troop commitments so far in Mozambique, where Exxon hopes for immense gas profits, the U.S. as of this writing has sent 60 troops. In central Africa, the U.S. African Command (Africom), has a more substantial presence.

An Africom mission statement reads,

1 Statement to House Armed Sercices Committee xxxx date by xxxx.

> U.S. Africa Command, with partners, counters transnational threats and malign actors, strengthens security forces, and responds to crises in order to advance U.S. national interests and promote regional security, stability, and prosperity.

The "prosperity" part for an African nation depends on U.S. corporations gaining resource-extraction contracts and taking their prosperity first. U.S. planning documents long have held that American armed forces are responsible for ensuring a steady stream of raw materials for corporations – especially energy corporations – and to maintain unimpeded movement of goods through shipping channels. An early example is *National Energy Policy* (May 2001) from Dick Cheney's National Energy Policy Development Group.

Some of AFRICOM's known permanent and semi-permanent military bases on the African continent, 2019.

The United States already has 29 known military facilities in 15 countries on the African continent, while France has bases in 10 countries.

Africom is headquartered officially in Stuttgart, Germany (on World War II-era U.S. military-occupied property), but Africans concerned with national sovereignty fear Africom will move or already has covertly moved its command post to African soil.

> Since it was set up in 2007, the U.S. government's Africa Command (AFRICOM) has not been able to find a home on the African con-

> tinent; The African people continue to pressure their governments not to give in to U.S. demands to shift the AFRICOM headquarters from Europe to Africa.[2]

A rumor holds that a new super-Embassy compound on 10 acres in the African nation of Eswatini houses a secret Africom headquarters.

By 2014, the sub-Saharan region, Sahel, had experienced a number of conflicts, driven variously by militancy, by piracy, or by smuggling. Using the pretext that these activities threatened national security for France and for the United States, these countries intervened militarily across the Sahel. This continued a colonialist tradition begun in the 19th century when European powers invaded the Congo to extract copper from the Katanga region. Katanga also has deposits of cobalt, zinc, coal, silver, and cadmium.

In 2007, a large discovery of oil, estimated at 1.7 billion barrels, was made at the border of Democratic Republic of the Congo and Uganda, on the shore of Lake Albert in the Kivu region. Congo and nine surrounding countries had barely emerged from years of civil wars funded by conflict minerals.

But unsurprisingly, weapons flowed to the oil-rich North Kivu region, and conflict continued, between a Ugandan rebel group called the Allied Democratic Forces (ADF), which has operated in the Congo since the early 1990s, and the Congolese military. In August 2007, the Congolese army attacked an oil barge near the disputed Rukwanzi Island on Lake Albert, killing a British oil surveyor contracted by companies exploring for oil in the area.

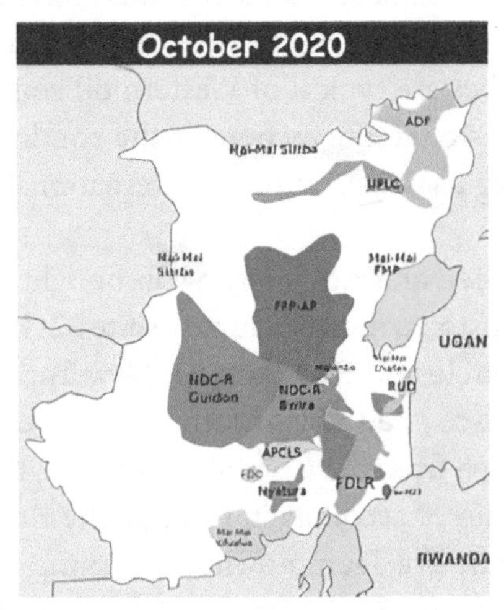

Approximate map of current military situation in Kivu.

2 Dossier No. 42: Defending our sovereignty: U.S. military bases in Africa and the future of African unity," Monthly Review Online, July 6, 2021

In 2012, the think tank International Crisis Group warned,

> An upsurge in fighting since the start of 2012, including the emergence of a new rebellion in North Kivu and the resumption of armed groups' territorial expansion, has further complicated stability in the east, which is the new focus for oil exploration. New oil reserves could also create new centres of power and question Katanga's (DRC's traditional economic hub) political influence. Preventive action is needed to turn a real threat to stability into a genuine development opportunity.[3]

ICG is a think tank founded by America's old-wealth oil and mining families – Carnegie, Rockefeller, and Koch. It is based, fittingly, in Brussels, Belgium, home of the Unione Miniere company that promoted Katanga's secession from the Congo – to secure copper resources for the company – in 1960.

"Preventive action" is code for putting U.S. soldiers in place of African soldiers, whom the countries' rulers typically, and deliberately, underpay and undertrain, in order to lessen the chance of a well-organized coup d'état.[4]

And, just while the ICG think tank was warning that "preventive action" was needed in Congo, the "gatekeeper to Congo's oil wealth," Denis-Christel Sassou Nguesso, son of the Congolese leader, was buying a luxury condo in Miami. Beginning in 2009, Sassou Nguesso transferred about $10.3 million to bank accounts in South Florida, a U.S. federal suit says, alleging he solicited bribes and stole millions in oil revenue. As we have seen, such bribes are typical of Western oil companies. U.S. prosecutors charged that stolen money bought the condo, and described the actions of Sassou Nguesso as part of an "international money laundering conspiracy."

Then, in 2014, the Nguesso family again bought expensive U.S. real estate with allegedly stolen money – luxury suite 32G in the Trump building at Columbus Circle and Park Avenue, New York City. Investigators following money discovered that shortly after large amounts of public funds were withdrawn from the Congolese treasury, that money was funneled through a chain of accounts in Cyprus, the British Virgin Islands, and Delaware. At least five Trump-owned or Trump-led companies were involved in the sale of unit 32G, according to documents reviewed by the NGO Global Witness.

3 ICG Web site, July 12, 2012

4 "The Bewildering Search for the Islamic State in Congo," The Nation, April 20, 2021

In the DRC during coming years, President Denis Sassou Nguesso, father of Denis-Christel, presided over events that provided a pretext for U.S. military involvement in the country.

In January 2021, a delegation of Africom officers arrived in the Nguesso-ruled Congo for discussions with the Congolese military. They proclaimed that the DRC urgently needed to agree to the following:

> Cooperation and engagements, security and stability efforts, and working together to further professionalise the DRC military and strengthen ties.

Following up on this proclamation on March 10 2021, the U.S. State Department made specific allowance for troop deploy to Congo, by designating the Allied Democratic Forces (ADF) as a "Foreign Terrorist Organization" composed of "Specially Designated Global Terrorists." There is no evidence to link the ADF to the Islamic State, according to local organizations and the UN Group of Experts on the DRC. Rather, the State Department based its decision on a claim made by an affiliate of the Texas-based hedge fund Bridgeway Capital Management.[5] The main area for troop deployment will be adjacent to the oil reserves.

As one African scholar put it,

> The U.S. military will also continue to provide stability for the African strongmen, who have come to rely on U.S. support for their longevity.[6]

The U.S. inserts troops to Africa whenever an ostensible reason presents itself – even stifling a pro-democracy movement if necessary – in Eswatini, the former Swaziland – going directly counter to announced U.S. policy, with no apology.

In Eswatini, at a huge new U.S. Embassy complex of nearly 10 acres was built starting in 2013. It was quickly rumored to be covert headquarters for Africom. Currently, U.S. troops are deployed in Eswatini to enforce a state of martial law declared by the country's absolute monarch, King Mswati III(Mswati's politics are such that he chooses to not recognize the Beijing government).

King Mswati III

5 "The Bewildering Search for the Islamic State in Congo," The Nation, April 20, 2021

6 Ibid to 9.

Demonstrations in the country began as a drive for trade unionism. Beginning in 2011, Mswati's riot police resorted to teargas, water cannons, and beatings of demonstrators while arresting union leaders, activists, and journalists.

In a notable coincidence in February 2013, just as the popular democratic movement was gathering strength (after Mswati announced his government would not accept any further "petitions" from citizens), the U.S. began construction on its super-Embassy complex.

Perhaps as a result, conflict escalated from there in Eswatini. Journalists say that doctors have confirmed at least 50 deaths at the hands of security forces. A current round of protests leading to U.S. troop deployment began in May 2021 when a law student was murdered "in circumstances that suggested police involvement." And correspondingly, demonstrators' demands have escalated as well. Far beyond recognition of trade unions, demonstrators now call for full dissolution of the Mswati III monarchy – in favor of democratic elections.

So, the picture now is one of U.S. Marines lined up against pro-democracy demonstrators, outside aUS Embassy – *Marines who were not deployed until protests called for full electoral democracy.* This is quite remarkable, of course, because the U.S. routinely has supported so-called "pro-democracy" movements.

What is different here? No clear answer has been offered.

The Marines who deployed on June 30, 2021 were sent overseas at the request of the State Department's Diplomatic Security Service, the department's law enforcement and security arm. This agency is tasked with "securing diplomacy" and "protecting the integrity of U.S. travel documents."

U.S. Marines assigned to Marine Security Augmentation Unit (MSAU), walk up range during a weapons field test of the M27 Infantry Automatic Rifle at range one, Marine Corps Base Quantico, Va., Feb. 21, 2014. (U.S. Marine Corps photo/ Staff Sgt. Ezekiel R. Kitandwe)

Sahel – "The Shore of the Desert"

Over the decades, Burkina Faso's desert areas have been spreading. Colonialism, coups, corruption, poverty and ethnic strife have all taken a toll. Life is not easy, but existence of a democratic, multi-party system of government in Burkina Faso had kept unrest to a minimum.

As late as 2013, a State Department report noted,

> There were no recorded terrorist incidents in Burkina Faso, which is not a source for violent extremist organization recruitment efforts or home to radical religious extremists.

Indeed, when a rebellion occurred the next year, 2014, it was pro-democratic – it was a rebellion against a constitutional amendment proposed to allow President Blaise Compaoré to run again, and extend his 27 years in office.

When Jihad began in Sahel around 2015, the U.S. took a rigid, militarist stance – i.e., social conditions did not matter – because the situation supposedly was simply one of aggressive Islam versus the West. This stance raised doubts immediately among those informed about African societies, including former U.S. military officer Danny Sjursen, who soon began writing about it. In blaming Sahel conflicts on the U.S.'s formation in 2009 of a "*Trans-Sahara Counterterrorism Partnership* (TSCTP), Sjursen has written,

> *The main match was lit in 2009, when Burkina Faso joined the Trans Sahara Counterterrorism Partnership (TSCTP),* a joint State-Pentagon, but military-skewed, slush fund for training, advising, and equipping local regional security forces to counter negligible, if not nonexistent, terror. The core problem was philosophical – of America imposing, and Burkinabe political elites willingly applying, a counterterrorism formula that didn't address, and actually inflamed, the long-neglected nation's foundational cornucopia of conflict kindling.[7]

Even the mainstream *New York Times Magazine* wrote,[8]

> Although American commandos continue shuttling into the (Burkina Faso) borderlands, the United States has found itself unable to effectively train, arm and support local security forces with-

7 *LA Progressive*, June 19, 2021

8 "How One of the Most Stable Nations in West Africa Descended Into Mayhem," *New York Times Magazine*, October 15, 2020

> out contributing to the conditions that push locals into the arms of jihadists.

These conditions, not spelled out by the *Times Magazine,* included persistent government impunity, repression, and challenges of governance, especially in the domain of police, defense, and justice affairs.[9]

The *Times Magazine* continued,

> Burkina Faso has become just another of many countries – from Afghanistan to Iraq, Libya to Somalia – where the United States has spent time and energy and money, only to see the mission stagnate, worsen, or outright fail.

U.S. troops essentially just continued the habits of the Burkina military – repressing dissent in general and Muslims in particular. As one Africa expert notes,

> Militaries in the Sahel are predominantly non-professional. Since independence, they have rarely fought against other states, instead being used largely for the internal repression of political opponents. International support always risks providing a security net that deters the military and the ruling class from reforming governance.[10]

He cites Stampnitzky, L., "Disciplining terror: how experts invented 'terrorism,'" Cambridge: Cambridge University Press; 2013.

China

Much has been made of the amount of Chinese investment in Africa, which amount is now around $153 billion. Indeed, a full 20 percent of African government external debt is owed to China. But 32 percent of African government external debt is owed to private lenders, and 35 percent is owed to multilateral institutions such as the World Bank.[11]

China's lending to Africa is shrinking, and, in rate, it has dropped below that of the U.S., as is shown by the following graphic on the next page:[12]

As close observers have noted,

> Chinese financiers have committed $153 billion to African public sector borrowers between 2000 and 2019. After rapid growth

9 ChathamHouse.org Web site, March 2,2021

10 Marc-Antoine Pérouse de Montclos, "Rethinking the response to jihadist groups across the Sahel," Chatham House, March 2021 . Dr. Perouse de Montclos is a political scientist and a senior researcher at the Institut de recherche pour le développement (IRD), Paris.

11 JubileeDebtCampaign Web site

12 CarnegieEndowment.org Web site, June 2, 2021

in the 2000s, annual lending commitments to Africa peaked in 2013, the year the Belt and Road Initiative was launched. By 2019, though, new Chinese loan commitments amounted to only $7 billion to the continent, down 30 percent from $9.9 billion in 2018."[13]

FIGURE 1
China's Lending Is Large But Shrinking

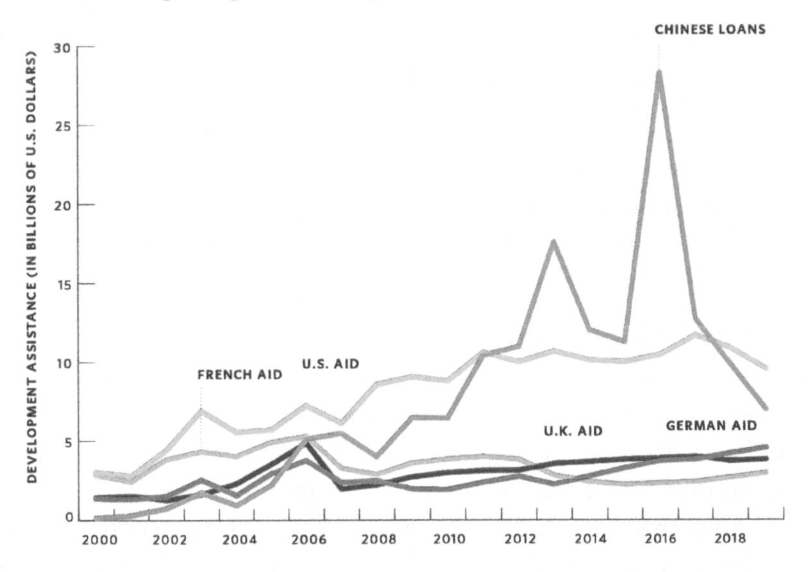

SOURCE: Author's calculations from Chinese Loans to Africa Database, Version 2.0, published by the China Africa Research Initiative and Boston University Global Development Policy Center (2021), retrieved from https://chinaafricaloandata.bu.edu; and "Total Official Development Flows by country and region (ODF)," OECD.Stat, https://stats.oecd.org, retrieved on May 10, 2021.

Nevertheless, a narrative has arisen of "China debt trap for African nations." This appears to have been to obscure the fact of Western debt-trapping of African nations.

A debt trap forces a country, in exchange for money, to modify its economic practices to suit the desires of a lender. Lenders in this position have been described as predators.

In Mozambique, loans that preyed on the nation – in the MozamScam described in Chapter One – were arranged by an international spy-based scheme.

And even a private transnational Western corporation, Glencore, has preyed financially – on the African nation of Chad (Glencore is one of the palm-oil traders that funded the Colombian death squad AUC, cf. Chapter 5). The loan to Chad by Glencore and other creditors was a "cash-for-crude" deal, in which imposed conditions forced Chad to encumber its oil assets toward repaying the loan. In fact, a majority of the country's oil

13 Ibid.

revenues from its oil sector have gone primarily to repaying Glencore.[14] This appropriation of revenues by foreigners meant domestic social programs in Chad could not be funded adequately.

The Western-controlled International Monetary Fund (IMF) and World Bank recently pressured Glencore to renegotiate the $1 billion loan to Chad.

But, these agencies have themselves been accused of predatory lending to Africa. In fact, the IMF states, baldly but forthrightly,

> When a country borrows from the IMF, its government agrees to adjust its economic policies to overcome the problems that led it to seek financial aid. These policy adjustments are conditions for IMF loans and serve to ensure that the country will be able to repay the IMF.

A typical IMF loan condition is that a country "eliminate price controls." This tends to produce inflation, especially in a conflict-ridden country. "No-price-controls" allows whoever owns the means of production to take extra profit.

All this may serve ultimately to "ensure" repayment, or it may not. The record shows that typically, "no-price-controls" tends to ensure continuing income inequality in a nation; no reform comes as long as that nation is beholden to the wealthy, foreign and domestic, because they generate money to repay the IMF.

Chinese Investment Is Said to Be Differentt

Two senior U.S. professors published an article in *The Atlantic* in February 2021 with the telling title "The Chinese 'Debt Trap' is a Myth." They found that the charge of "colonialism" against China is made, chiefly and hypocritically, by countries that themselves have a perfectly well-documented history on their own of colonialism and neo-colonialism in Africa (that is, they found that the narrative on investment in Africa is controlled by Western writers). As researchers at Carnegie Endowments found on the subject,

> There is a conspicuous absence of African scholars' analyses from a global debate that can only be enriched by their voices and lived experiences. The systemic exclusion of African voices in global centers of knowledge production persists.[15]

14 ExtractiveIndustriesTransparencyInitiative.org Web site February 15, 2021

15 Ibid. to 20

Long-time Africa scholar Vijay Prashad, writing for the Tricontinental Institute for Social Research, observes,

> Chinese aid (to Africa) – unlike IMF aid, Western commercial investment, and overseas development assistance – does not come with the vice of debilitating conditionalities.
>
> Evidence for more favourable terms comes in the various agreements signed by China, but more than that, it comes from China's theory of patient capital, which has until now been adopted within the boundaries of China but has slowly–through Chinese state banks–emerged as a major investor outside its territory.[16]

And significantly, China regulates cryptocurrency – strictly, both in an attempt to prevent currency flight from the country and to rein in organized crime. In May 2021, China banned financial institutions and payment companies from providing crypto-related services. Then, a month later police in China arrested more than 1,100 people suspected of using cryptocurrencies to launder illegal proceeds from telephone and Internet scams.[17]

Ironically, China has made hundreds of arrests of people that highly likely are among the crowd that was scamming users of Chevron bitcoin ATMs. The nation's Ministry of Security announced in June 2021 that

> • Police in China arrested over 1,100 people suspected of using cryptocurrencies to launder illegal proceeds from telephone and Internet scams in a recent crackdown.
>
> • The arrests came as authorities in China step up their crackdown on cryptocurrency trading.
>
> • The money launderers charged their criminal clients a commission of 1.5% to 5% to convert illegal proceeds into virtual currencies via crypto exchange.

War With China?

A U.S.-China war is predicted in some circles. The prediction refers to a perceived or at least stated need of the U.S. to "deter competition" from China.

The Pentagon's 2022 Defense Budget Overview reads in part,

> "China poses the greatest long-term challenge to the United States, and strengthening deterrence against China will require the Department of Defense to work with other instruments of national

16 Ibid. to 9
17 CNBC, July 7, 2021

> power. A combat-credible Joint Force will underpin a whole-of-nation approach to competition and ensure the (U.S.) leads from a position of strength."

Reportedly, a significant U.S.-China dispute exists over right of U.S. warships to navigate the South China Sea. As such, the mainstream media in effect predicts that if a war with China comes, it will be a naval war.

A U.S. Navy destroyer-class ship launches a guided missile during "Rim of the Pacific" military exercises in the Pacific Ocean on August 6, 2020.

However, the record suggests that between the U.S. and China, land war is more likely – in Africa. Attractively in Africa, neither government would fight on its own soil. And there, mines and oil wells would be battle prizes (just as in current rebellion conflicts). As such, each warring nation would be practically committed not to destroy mines, but simply to retake them after losing them. As with the local rebellions, such a U.S.-China land conflict likely would become prolonged – as a profit source for the military-industrial complex of each country.

Chapter Fourteen

The West Can't Fix What It Is Breaking in Africa

Innovation is much praised in our modern hyper-technological culture. A theme around this is the notion that we can innovate our way out of anything. Whether or not this is simply foolish remains to be seen, but the record is full of cases of unintended consequences flowing from highly touted innovations in technology, such as, for example, social media. This technology was supposed to innovate our way out of societal fragmentation, atomization, loneliness. But sometimes a fix breaks things, putting us in a worse "fix," and then is unable to fix that.

Western Banking's Innovative Foible – CryptoCurrency

A new version of Western capitalism/consumer culture is catching on in Africa – electronic-currency investments and the swinger culture that, the record shows, goes with them. Tanzania, with socialist leader John Magufuli replaced by Samia Suhulu Hassan, now is at the forefront of this.

In November 2019, the Bank of Tanzania cited local law as a reason for banning digital assets. Now, the central bank says the ban is poised to be reversed following pro-crypto comments from President Samia Suluhu Hassan:

> In the financial sector, we have witnessed the emergence of blockchain technology or cryptocurrency. Many countries in the world have not accepted or started using these currencies. However, I would like to advise the central bank to start working on those issues. Just be prepared.[1]

Suhulu Hassan's decision reflects a desire for her nation to not be left out of a technological trend that, all else being equal, could benefit some economic sectors in the country. However, all else is far from equal. Cryptocurrency/blockchain/bitcoin is a standard Western technological innovation – i.e., first, government has not regulated it, and second, people

1 The Daily Hodl Web site June 27, 2021

have scrambled to adopt it – *all ignoring the possibility of unintended consequences.* And for bitcoin, an unintended consequence looms so large as to tax credulity. The record suggests strongly that if bitcoin conquers Africa, the planet will not survive.

The reason for this is that use of cryptocurrency drastically speeds global warming – by consuming surprising, no, mind-boggling, amounts of energy. This will be reviewed below.

And unsurprisingly, the big complex of deliberate global-warming industries – Big Oil, organized crime, and Big Real Estate – is a champion of cryptocurrency. Chevron and Total recently joined a blockchain-based electronic commodities-trading "platform"[2] that was started, in 2017, by a consortium of oil majors BP Royal Dutch Shell, and Equinor alongside energy traders Koch Supply and Trading and Mercuria Energy Group, based in Houston, Texas.[3]

Chevron, Bitcoin ATMs, and Organized Crime

Around 2020, Chevron began installing bitcoin ATMs at its gas stations. Soon, the machines began displaying this message:

> WARNING! Have you been suggested a job opportunity and then asked to send bitcoins using ATM? Or found a great deal, e.g. car on craiglist, and was asked to pay in bitcoins using ATM? You are highly likely a victim of a scam.

Investigators soon turned up evidence that, sure enough, bitcoin ATMs at Chevron stations and elsewhere were involved in crimes. In a press release titled, "Bitcoin ATMS: Scams, Suspicious Transactions, and Questionable Practices at Cryptocurrency Kiosks," the State of New Jersey Commission of Investigation reported,[4]

> The Commission uncovered numerous instances where unwitting victims were duped into sending cryptocurrency to unknown wallets through the machines.

This was in February 2021.

Two months later, in April 2021 as the fraud was spreading, Exxon began opening bitcoin ATMs at its gas stations. The timing makes this move questionable – what was Exxon thinking? Did it not care?

2 Vakt is its name.
3 Reuters, January 16, 2019
4 In *February 2021*

It arguably should have cared, because the quickly spreading fraud in oil-company bitcoin ATMs was evidently the work of organized crime, quite likely friends or "friends of friends" of those hackers who penetrated the Citi ATMs at 7-Eleven stores – while a Citi-controlled trust was buying hotels in Mexican towns that are run by criminal cartels, as discussed in Chapter 7.

Gangland's entry into bitcoin was and is a large contribution to global warming. This will be reviewed below.

How does bitcoin accelerate global warming?

The Digiconomist's Bitcoin Energy Consumption Index estimated that *one bitcoin transaction takes 1,544 kilowatt hours of electricity to complete.* Of course, in generating this amount of electricity, coal, gas, and oil are burned and thus emit tremendous amounts of carbon dioxide. According to U.S. Energy Information Administration calculations, generating just 1,000 kilowatt hours of electricity emits nearly 1,000 pounds of CO_2 to the atmosphere.

As such, the growth of bitcoin could produce enough emissions by itself to raise global temperatures by 2 degrees Celsius -- past the 1.5 degree-Celsius limit set by The Paris Agreement – and soon.[5] That's if, in Western style, it's left unchecked and is widely adopted globally with little or no regulation and no thought to unintended consequences.

As researchers wrote in a 2018 study *estimating that in 2017, Bitcoin usage emitted 69 metric tons of* CO_2*,*

> Bitcoin is a power-hungry cryptocurrency that, should it follow the rate of adoption of other broadly adopted technologies, could alone produce enough CO_2 emissions to push warming above 2 degrees Celsius within less than three decades.

The lure of easy money – wealth there for the taking – has led to use of the metaphor "mining" to describe a bitcoin process.

Digiconomist estimates that electronic bitcoin "mining," the process by which a participant possibly can generate or capture bitcoins for himself, accounts for 0.29 percent of the world's annual electricity consumption. The mining of a single bitcoin block—a block of transaction data on the bitcoin network—consumes enough energy to power more than 28 U.S. homes for a day. A user must verify an entire megabyte worth of *transactions* merely to become eligible to earn *bitcoin (by using his*

5 "Bitcoin emissions alone could push global warming above 2°C," *Nature Climate.*

computer further to solve computational puzzles) —not *everyone* who verifies an entire megabyte of *transactions* gets paid out.

So, for bitcoin pursuit, "gambling" is just as apt a metaphor as is "mining." The mining metaphor is most apt in expressing that like physical mining, cryptocurrency "mining" consumes enormous amounts of energy and speeds global warming to a degree that the public is not yet entirely aware of.

Petroleum Flaring and Bitcoin

Exxon has proposed to collect petroleum that otherwise would have been burned off into the atmosphere in "flaring" and sell the petro through the medium of bitcoins.

As of this writing, the current Biden Administration proposes some further regulation of bitcoin, with Secretary of the Treasury Janet Yellen pointing out the industry's propensity for money-laundering. However, the pressing national issue is the global-warming acceleration caused by the thousands of bitcoin money launderers with their thousands of megabytes of electricity consumption. Secretary Yellen, EPA administrator Michael Regan, and President Biden deserve to hear this from millions of Americans who push for the Administration to go far in bitcoin regulation – before it is too late, for the planet.

Cryptocurrency and Organized Crime

Like online gambling, the cryptocurrency industry is quite inviting to organized crime, so much so that it has earned the sobriquet "La Crypto Nostra." Despite Forbes magazine's faint praise – "The majority of *cryptocurrency* is not used for criminal activity" – the crypto industry attracts such a large group of criminals that they, like the high-roller customer-base in Las Vegas, can get "comps" – special treatment.

The U.S. Justice Department reported,[6]

> Rossen Iossifov designed his business to cater to criminal enterprises – by, for instance, providing more favorable exchange rates to members of (a Romanian criminal) network. Iossifov also allowed his criminal clients to conduct cryptocurrency exchanges for cash without requiring any identification or documentation to show the source of funds, despite his representations to the contrary to the major bitcoin exchanges that supported his business.

6 January 12, 2021

In January 2021 Iossifov was sentenced to prison for money laundering in "a transnational and multimillion dollar scheme."

In 2020 alone, criminals laundered $1.3 billion through a total of 270 cryptocurrency addresses.[7] As such, organized crypto-crime was responsible for a notably consequential amount of carbon-dioxide emissions to the atmosphere, in a pattern that has highly likely existed for years, ever since bitcoin emerged, in 2009.

As of this writing, the current Biden Administration has shown signs it will oppose bitcoin.

Kenya

Kenya is a major ally of the U.S. in Africa. It has lots of oil.

In 2019, Kenya ranked 137 out of 198 countries on Transparency International's Corruption Perceptions Index. By around 2017, abuse of power by Kenyan police became so egregious that the State Department's Bureau of International Narcotics and Law Enforcement Affairs (INL) announced it would "provide support" by pairing experienced U.S. police trainers with Kenyan counterparts and "creating a customized police oversight investigator training curriculum."

To the extent this was aimed at more than sanitizing publicly a close U.S. relation with a dishonest government and its brutal police force, the move had little chance of working. A s in several other African nations, the public police in Kenya are allowed to contract with private security firms, which number around 2,000 in Kenya and dominate law enforcement. Law enforcement is primarily for the wealthy, who can afford to pay. As such, although Kenya maintains a 1-to-400 ratio of public officers to citizens, it has become known as "over-policed and under-secured." By 2020, Kenyan police were well-documented, on numerous occasions to have fired live ammunition at non-violent anti-government demonstrators.

In terms of volume of mobile money transactions, East Africa dominates the global market. The region is No. 1 in the world,[8] which remarkable status is mainly due to Kenya, where the number of mobile-money subscriptions rose to 29.1 million during 2021.

Correspondingly, with an online gambling industry benefiting from mobile money, Kenya is the third largest online betting market in Africa, behind South Africa and Nigeria. Kenyans spend around $50 per month on gambling, and 96 percent do that via mobile phones. In 2021, the gov-

7 Al Jazeera, February 12, 2021

ernment tapped the smart-phone gambling cash cow with a 1.5 percent tax on digital services provided.

In Kenya, much armed conflict is the result of soldiers intervening in disputes over resources. Resource-extraction industries are largely responsible for this. Some armed conflict in Kenya occurs between the military and fighters from al Shabaab ("The Youth"). Al Shabaab is loosely based in Kenya's neighbor to the east, Somalia.

Kenya's interest is strategic and economic: a semi-autonomous Juba state as a buffer-zone from Al-Shabaab attacks both on its tourism industry, and a massive port development project at the city of Lamu; a secure access to the Kismayo market; and an influence over oil and gas deposits in a contested maritime zone.

In its early ambition, the Chinese-built Lamu port figured as connecting the landlocked East African economies to global trade routes. More specifically, it was envisioned as an alternative outlet for South Sudan's oil, which is currently pumped via the Greater Nile Oil Pipeline to Port Sudan.

The project's new aim, distinctly Western-style and Western-oriented in that it claims vast tracts of land for development, calls for multiple resort cities, airports, numerous industrial areas, an oil pipeline, and a damming of the Tana River. Domestic oil is believed to be present offshore of Lamu. So the future holds yet another Western "Petroleum City" on Africa's east coast.

The effect will be complete with a vast increase in oceangoing shipping and multiple airports along the Lamu corridor. So far, around 19 shipping lines have inspected the Lamu port. The Kenya Ports Authority advertises generous promotional offers to shipping lines..

Global shipping is the world's sixth highest emitter of carbon dioxide, producing 2.4 percent of global greenhouse gas emissions. In 2015, rather than seeking to comply with the Paris Climate Agreement, the global shipping industry simply decided not to abide by the international agreement at all.[8]

This will have a consequential effect on accelerated global warming.

In the last 15 years or so, Lamu has become a highly volatile region. Attacks by the al-Shabaab militant group have brought violence to the area and turned it into a highly securitized region. Security operations have significantly reduced incidents, but periodic al-Shabaab attacks have affected construction activities. Just as the civil war in Mozambique reflects dispute over benefits from petroleum generated at its coming "Gas City,"

8 Bloomberg, December 10, 2015

the Lamu-area project is likely to generate an armed conflict of its own, further changing the planetary climate.

A Technological Fix Is Not Achievable

The crisis of global warming has given rise to a "green-energy revolution." However, this revolution is expected largely to be carried by industries, and as a result, this "revolution" is expected to drastically increase the demand for the mined metals used toward renewable energy (such as long-life batteries, and wind turbines).

That is, "renewable energy" will depend on mining. It will depend on significant mining of "rare earths" (a set of 17 nearly indistinguishable heavy metals). Of these, China has 50 percent of global reserves and accounts for 90 percent of production of rare earths. The green revolution also calls for large-scale mining of cobalt and lithium.

Concerning mining of rare earth minerals, the mining industry recently began eyeing Africa.

As the journal *Mining Technology* blithely puts it,[9]

> Cashing in on deposits of rare earth materials is a savvy way to secure a seat at the table of the blossoming green economy.

Concerning lithium, a major producer by 2022 figures to be the Democratic Republic of the Congo, where the government has fast-tracked beginning lithium production..[10]

In Tanzania's first rare earths mine, Peak Resources' Ngualla project will exploit one of the largest deposits in the world of a rare earth called neodymium-praseodymium.

The journal *Mining Technology* enthused,

> With the government hand-over (to new President Samia Suhulu Hassan), a new emphasis on mining has taken the driving seat, with the Ministry of Minerals reportedly turning its attention to boosting mining's contribution to the GDP. The new administration is also seeking to enact mining companies' right to mine, as well as Tanzania's right to benefit from its mineral wealth.

The journal continued,

> "A Bloomberg report found that the government intends to increase mineral earnings by 33 percent over the next three years, marking

9 July 14, 2021

10 TheAfricaReport Web site, June 22, 2021

> a revenue increase of up to $302 million between 2023 and 2024. With such a setup, the foundations have been established for Tanzania to see a boom in its mining fortunes, and Peak is in prime position to ride the wave.

As we saw earlier, resources mining in African nations is a source of prolonged armed conflicts that accelerate global warming. So, turning to industry to solve the global-warming problem is simply pulling down carbon emissions with one hand while pushing them up with the other.

Chapter Fifteen

Scenarios To Come

It is likely that in the U.S., no matter what "green economy" might arise, significant fossil-fuels extraction will continue (the Texas Lobby is that strong). In other countries, things will be different.

After Oil, What?

Angola faces what is called the "structural decline of its oil production." This reflects the deep involvement Angola had with dominating Western oil companies, who have built massive physical plants that need maintenance. Under an "oilfield production-sharing agreement" (PSA), shareholders (in this case, the Angolan government) are required to pay the field's operator (in this case, a Western oil company) to cover the costs of maintaining and expanding production. This stricture is now enabling a Western "privatization," partial and piece by piece, of Angola's public oil company Sonangol, a feat by industry nearly matching the successful 100-percent capitalization of Mexico's Pemex.

On Wall Street, it was big news when an oil company executive told Reuters,[1]

> Sonangol hasn't paid cash calls for a while.

When the oil exec spoke, the plan was already a fait accompli for a partial, over-time privatization of Sonangol. He said that debts on just one of the eight fields for which the company offered partial stakes ran above $100 million. As a test for the extreme move of auctioning public-oil shares, Angola is auctioning shares in its state bank and around 20 percent of other state-owned assets – in a sale expected to be complete by the end of 2022.[2]

On oil PSAs with Western companies, Sonangol said the company's financial commitments through 2027 would run up to $7 billion for developing fields, maintenance, debt servicing, and cash calls. Perversely, it is unlikely Angola has enough oil left to pay this debt.

---

1 June 30, 2021
2 Bloomberg, April 19, 2019

In hindsight, oil production-sharing agreements should never have been signed that would prove so ruinous for Angola. The agreements were signed by the same run of dishonest public officials described in Chapter 4 as having been bribed lavishly by Western companies including Dick Cheney's Halliburton.[3]

And not surprisingly, a raft of U.S. companies are planning to profit from Angola's desperate situation, its inevitable "diversification" away from oil. The U.S. Chamber of Commerce has opened an affiliate in Angola ("AmCham Angola"). Veira de Almeida (VdA), an investment-law firm, founded AmCham Angola along with Exxon, Chevron, BP, and other large corporations.[4]

Tourism and with it gambling are being pushed as Angola's path beyond its oil-field maintenance debt. The country now has a Ministry of Tourism and Gaming, one of several actors touting the plan as an "essential source of economic development for Angola." Some of these touts are quite loud.

> "Angola plans to become the Macau of Africa," according to a publication whose goal is "bringing together thousands of investors and innovators in Blockchain, Fintech, Artificial Intelligence, Quantum Technology, Big Data, and the Internet of Things.[5]

Angola currently has around seven casinos. Each of the planet's approximately 2,150 casinos is a heavy user of electricity. If Angola is to become the Macau of Africa simply to pay off debt to Western oil companies, the thought is sobering. According to one report, Macau's 41 casinos consume some 467 million kilowatt hours per year of electricity. According to U.S. Energy Information Administration calculations, generating just 1,000 kilowatt hours of electricity emits nearly 1,000 pounds of CO_2 to the atmosphere.

Even if Angola were to top out at just 20 casinos, the industry in Angola would cause emission of more than 200,000 pounds of CO_2.

And that's without the exterior lighting for the casinos. The MGM casino and its brethren on The Strip in Las Vegas each uses approximately 800,000 kilowatt hours of electricity per year, and their exterior lightings consume a similarly gargantuan amount of energy. The Las Vegas Strip is so bright that it can be seen from space.[6]

3 Cf. New York Post, July 27, 2017
4 VdA Web site
5 Sigma Web site, July 15, 2020
6 As documented by astronauts, such as Canada's Chris Hadfield.

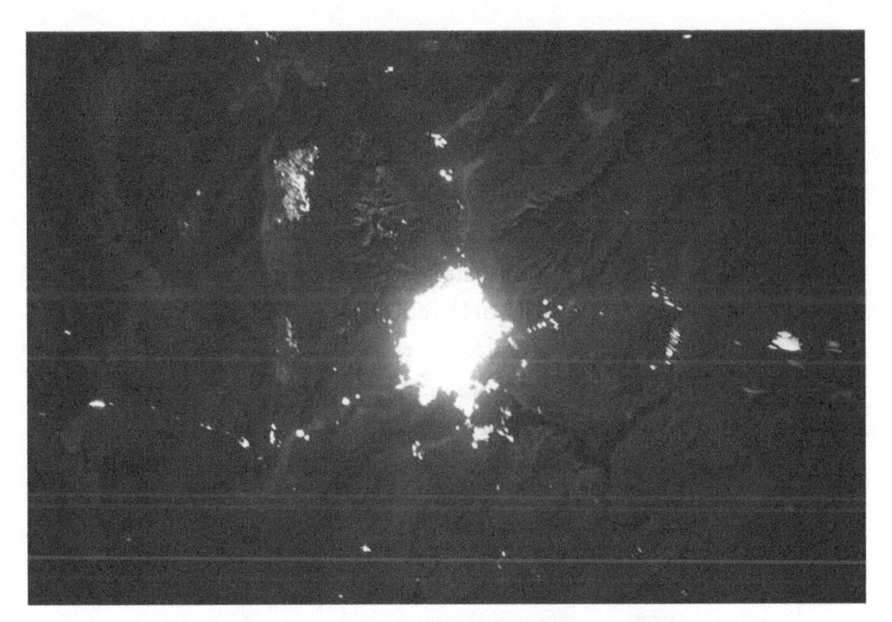

The Strip in Las Vegas as photographed from space.

Casinos on land, casinos on-line accessed by smart-phones, the mining of minerals used in those smart-phones, even the mining of minerals used in renewable-energy technologies. All of these cause and prolong armed conflicts in Africa, to an extent that is consequential for the acceleration of global warming (as said before, this acceleration is caused by deliberate, concerted actions of a complex of extractive and land-exploitive industries, and the point is utterly moot whether the deliberate actors consciously intend to accelerate global warming).

Wars and Warming

> *The International Committee of the Red Cross has found that of the 20 countries deemed most vulnerable to climate change, 12 harbor prolonged armed conflicts.*[7]

Use of explosive weapons in populated areas causes great destruction. This leaves a legacy of carbon emissions from trucks moving debris and from construction sites to rebuild buildings. Construction activity s a huge global warmer, emitting 3,560 metric tons of CO_2 globally in 2019. Additionally in conflict and post-conflict areas, deforestation is the norm. The Red Cross has estimated that in 2020, deforestation in tons of CO_2 – nearly four times the total emissions from the UK in 2020.[8]

7 Red Cross, July 9, 2020

8 Ibid.

Sites for extraction of oil often are targets of fighting. Explosions of weaponry at storage tanks and vehicle garages cause oil spills and oil fires that emit CO_2. This has been notable, for example, in the case of Colombia.

At the same time in conflict zones, environmental governance collapses. This creates and sustains impunity for violators of whatever emissions regulations may exist. Impunity extends into the post-war period, when weakened institutions and ungoverned spaces give free rein for global-warming activities to proliferate.

In war zones and post-war zones, after energy infrastructure and markets have been ruined, a need for fuel remains. Forests are cut – so that people can make charcoal. This is well-documented in Democratic Republic of the Congo.

Additionally, people in war zones may resort to oil refining of the artisanal (amateur) kind. This practice emits tremendous amounts of CO_2.

A makeshift oil refinery in a Syrian war zone

The more an armed conflict is prolonged, the more sources of anthropogenic emissions are locked in. This is caused by absence in wartime of external investment and weakened governance that might otherwise have replaced outdated, global-warming technologies.

An example of this is the practice of flaring – burning excess petroleum gas as a by-product of oil production, which releases it as CO_2. The volumes involved are huge. Countries in war display increased flaring of petroleum.[9]

9 World Bank Blogs, June 14, 2019

Wars destroy carbon sinks – vegetation and soils – burning them or mutilating them so they no longer trap carbon.

A researcher at Conflict and Environment Observatory wrote,

> We estimate that the carbon emissions from vegetation fires in conflict areas in 2020 alone were approximately 1,456 megatons of CO_2 emissions,[4] (more than) half the UK's annual emissions.[10]

African countries are among the planet's most climate-vulnerable, but with war-weakened governance, a country is less likely to participate in international projects to address climate change. Few conflict-affected countries have provided any biennial updates on climate changes to the United Nations Framework Convention on Climate Change.

As discussed, U.S. militarization in Africa, done explicitly to "ensure a steady flow of resources" from Africa, creates and/or prolongs armed conflicts in African nations. As we saw in Chapter 1, natural-gas resources are causing a militarization in Mozambique.

Mozambique

As of this writing, in general the situation in Mozambique still is, as it was described in Chapter 1, a roil of foreign oil companies, spies, and insurgents in a civil war over benefits from Western-developed natural-gas resources.

This final section offers updated specifics on the state of affairs in Mozambique.

With oil companies such as Total having suspended operations in Cabo Delgado during the fighting, spies might be the most important part of the current mix in the area of the proposed Gas City.

South Africa's *City Press* reported on July 4, 2021,

> South African spooks were *caught* (by Mozambican troops) and left stranded for nine days in an operation targeting insurgents in *Cabo Delgado* at the height of the attacks in that (area) earlier this year. *City Press* has learnt that the four operatives had their passports and equipment, including drones, confiscated in Maputo and lost contact with the SSA.... The team only returned home safely following intervention from Security Minister Ayanda Didio, who learnt belatedly that the Mozambican security counterparts were not happy.

The South African spooks were in Cabo Delgado trying to access information on Mozambique government's efforts so far in the civil war.[11] Why?

10 Conflict and Environment Observatory, Web site, June 14, 2021
11 EnergyVoice Web site, July 21, 2021

In April 2021, a South Africa-centric "regional agency" with military powers, the Southern African Development Community (SADC), had announced it would deploy agency troops to Cabo Delgado – claiming fears that a so-called Islamic insurgency would "spread" from Mozambique to neighboring African countries. More than three months after the announcement in April 2021, no SADC troops had arrived in Mozambique. This calls into question the announced urgency of any so-called Islamic threat.

Instead of from SADC, with approval from Mozambique, from Rwanda a total of 700 soldiers and 300 police began arriving to Cabo Delgado in July 2021. The purpose was, Africa observer Joseph Hanlon wrote,

> Establishing a secure perimeter around the liquefied natural gas project there ... an offensive to clear the area around the gas projects (in which) Rwandan troops will be charged with holding positions once they are cleared.[12]

The deployment of Rwandan troops was made at the behest of French President Emmanuel Macron, who asked Mozambique President Nyusi to request Rwandan troops to secure France's multi-billion dollar investments in the liquefied gas in the area in a project by Total.

Concerning SADC troops, on just when or if Southern African Development Community troops would arrive, Joseph Hanlon cited several African publications in writing on July 13, 2021,

> There has been some confusion. The prior deployment of Rwandan troops clearly put South Africa's nose out of joint (because) SADC wanted a South African major general to command the brigade. Mozambique expressed a strong preference for a Zimbabwean commander. South Africa's Defense Minister Nosiview Mapisa-Noakula (said) it was 'unfortunate' that Rwanda had deployed its troops into Mozambique before SADC.

Confirming the notion of confusion but making an excuse for it, South Africa's News24 wrote that although official positions were changing day-to-day on deployment to Mozambique, this was justified by "unrest" in South Africa.

Under the headline "Unrest in SA likely behind delay of SADC mission to Mozambique," the paper wrote,[13]

12 Joseph Hanlon Club of Mozambique Web site, July 16, 2021, citing a report by the publication Cabo Legado on July 13, 2021.

13 July 16, 2021

> Soldiers have been deployed to help quell unrest in parts of South Africa, delaying the 15 July deployment (to Mozambique).
>
> The deployment will be placed on hold, given the scale of the operation requirements here in South Africa," said Pikkie Greeff, national secretary of the South African National Defence Union.
>
> Greeff said the code of conduct for the deployment, published on Wednesday, marked a shift in the position put forward by Defence Chief, Major General Siphiwe Sangweni, in a security cluster meeting on Tuesday.

This was disingenuous. The same day, it was reported that SADC troops were deployed in Eswatini regarding unrest in that country – a complement to the U.S. military deployment in Eswatini discussed above.

Under the headline "SADC mission in Eswatini while uncertainty clouds Mozambique deployment," the story began,

> A Southern African Development Community (SADC) "technical, fact finding mission" is currently in landlocked eSwatini investigating and analysing unrest, while the SADC mission to Mozambique struggles to get off the ground.

After all this, finally on July 21, 2021, arriving in a surplus U.S. Army C-130 transport plane, some troops from the South African Special Forces deployed to Mozambique.

South African Special Forces in Mozambique

The point is that foreign nations are more interested than is Mozambique in having foreign troops in Mozambique. An implicit goal of foreign intervention seems to be the propping up of the simplistic and increasingly discredited narrative of "Islamic insurgency" in Cabo Delgado. Also implicit is a "Christianity versus Islam" theme modernized to acknowledge that control of resources is just as important as control of souls.

On July 16, 2021, the saber-rattling U.S. publication *The Trumpet* (Philadelphia) wrote, under the headline "EU to establish military presence in Mozambique,"

> With the unstated but vital goal of defending a vast trove of natural gas, the European Union announced a new military mission to Mozambique on July 12. The menace of radical Islamic terrorism has afflicted the country for several years, but worsened in 2021, prompting Mozambique's government to call for an EU presence in the country. Mozambique joins the list of African countries that have welcomed European troops.
>
> Since 2017, Mozambique has been plagued by violence from the Islamic State and its affiliates.
>
> Much of what is going on in Africa today – from Mozambique to Mali – is ultimately a battle for control of vital resources between two opposing power blocs: European nations, strongly influenced by Germany, and radical Islamic proxies supported by Iran. The most important reason to watch Europe's clash with radical Islam in Africa is because this is exactly what was warned about in great detail, thousands of years ago, in the pages of the Bible.

Spies, Troops, and Gas for Fracking: Mozambique and Wyoming

Mozambique and Wyoming?

In Powell, Wyoming, a place Dick Cheney likes to address fundraisers, in 2019 a Mozambican entrepreneur – whose large company somewhat resembles Cheney's Halliburton – stayed for three days to learn about fracking, sponsored by a U.S. government agency.[14] Abreu Muhimua, CEO of Engineering Corporation Mozambique (Encom), told a reporter,

> We would like to see Wyoming's private sector doing business with the Mozambican private sector.

Muhimua's company is responsible for modernizing a now-tourist route to Mozambique's Gorongosa National Park area. The agency bring-

14 Powell, Wyo. *Tribune*, November 29, 2019

ing Muhimua to Powell almost certainly was USAID, still involved in the Westernization of the Gorongosa area. Fracking is still legal in the U.S. because of the "Halliburton Loophole" crafted into the Safe Drinking Water Act, an exemption for fracking that was the brainchild of Dick Cheney's 2001 Energy Task Force.

Muhimua's host in Powell was Wyoming Completion Technologies (WTC), which like Halliburton makes and sells "frac plugs" in several African nations. The Powell party was evidently to celebrate WTC's fracking-supply entry into Mozambique. So, despite or because the U.S. EPA in 2016 found evidence that fracking-effluent pits frequently pollute groundwater that is drunk by rural well owners, the U.S. government as it did that day in Powell is helping export U.S.-made technology for the high-global-warming practice of fracking to Mozambique.

At the same time, "relief" money from the U.S. and other Western countries is being spent in Mozambique to alleviate the effects of accelerated global warming – in a practice that has been called "Rich nations treating the symptoms but not the disease." *Research shows Mozambique is the second country worldwide most vulnerable to climate-change effects.*

Shortly before Muhimua visited Cheney Country, UN Secretary-General Antonio Guterres in Mozambique had called out Western nations for being soft on oil companies.[15] Beira in Mozambique had just been largely destroyed by global-warming-caused Cyclone Idai.

> "I don't want to see (large UN nations') money financing destruction (by cyclone) like what happened in Mozambique," Guterres said. "It's urgent for the larger nations to stop funding fossil fuels."

In Britain, such a stop was called just *after* the UK's credit agency had funded 1 million pounds for a gas project in Mozambique.[16] The environmental group Friends of the Earth took legal action, noting the deal appeared to violate Britain's commitments under the Paris Climate Agreement. After signing the agreement, the UK all in all committed more than 3.5 billion pounds toward tropical fossil-fuel projects.

Pope Francis pointed up Western duplicity in 2019 in Mozambique.

> "It seems that those who approach with the alleged desire to help have other interests," Pope Francis told a crowd in Maputo.[17]

On tropical deforestation, journalist Jon Lee Anderson wrote,

---

15 Agence France Presse, July 12, 2019
16 *Guardian*, August 12, 2020
17 Reuters, September 6, 2019

> In these places, all that stands in the way of destruction (of crucial carbon sinks) is the ability of a few thousand indigenous leaders to resist the enticements of consumer culture.[18]

This is an important assessment. Along the lines of a drug through a needle, consumer culture is now being delivered in Africa speedily and in high doses – through new "private cities."

The publication *AfricaIntelligence* reported,

> Libertarian financiers who want to build cities run by private companies are increasingly interested in Africa, where they hope to seduce national governments with their utopian projects.
>
> Based on the assumption of a shared longing for new urban spaces, these cities come with promises of impressive amenities and functioning systems that will enable the urban lifestyle most Western cities provide.

The "ability of a few indigenous (African) leaders to resist Westernization," crucial for African autonomy, is nearly equally important in a larger context. This book has argued that – because of deliberate actions by a Western-industrial complex bent on extracting raw materials from Africa, undermining socialism, and building casino hotels on coastal lands depopulated by oil operations or by Sea Level Rise – Westernization of Africa is practically synonymous with deliberate acceleration of global warming.

As such, African leaders need the help – however indirect – of concerned Westerners. Scrutiny and loud criticism are crucial on oil-industry corporations' actions, especially those of Exxon and Halliburton.

18 "Blood Gold in the Brazilian Rain Forest," *The New Yorker*, November 11, 2019

Conclusion

So, we live in a world characterized by the profit motive – characterized by scamming, by suspect philanthropy, by bribing, by industrial drug dealing and money washing, by death squads, by preventable oil spills, by false estimates of CO_2 emissions, by U.S. political suborning of foreign governments, by real-estate moguls helicoptering to disaster areas, by spies subverting socialist governments, and by foreign militarization of resource-rich nations. All these things are connected, and they are all accelerating global warming.

This is a daunting prospect. Monstrously daunting.

But even in such times, it is still the duty of young gallants[1] to save the fair defenseless –in this case, the fair Gaia, Earth – from depredation by the grotesque. This book has been about who that grotesque is.

Carbon trapping is necessary to win the climate struggle, according to the International Energy Agency. As such, something is needed to get government officials talking daily about carbon-trapping technology – and how it can be funded by government. Here is where a pointed message to legislators – "think and fund carbon-trapping" – can help. They deserve this heads-up.

Full citations to scientific evidence of industry underreporting of carbon emissions are provided in Chapter Eight of this book, footnotes 1 through 10.

A shareholder in an oil company could show executives the scientific evidence of underreporting of carbon emissions.. She could ask her friends, electronically, to do the same, where they invest. Her husband could write the Environmental Protection Agency, and similarly inform regulators of routinely underreported emissions. Additionally, he could signal the EPA about a new grass-roots push forming toward holding oil companies accountable – he could state flatly that repeated Climate Marches on Washington are the future.

The Women's March on Washington in 2017 is a perfect model for climate activism to force Congress, especially the Senate, to recognize

1 By definition, gallantry is something that can be shown, and it has been shown, by women and children as well as by men.

evidence, previously ignored or suppressed, that oil industry carbon emissions have been grossly underestimated. Organizers of that march included the following:

- The League of Women Voters
- The Natural Resources Defense Council,
- MoveOn.org,
- Oxfam, and
- Greenpeace USA.

Before such marches got stimulated, each U.S. Senator would have received messages containing citations of evidence on oil-industry underreporting of emissions and also containing warning that in forthcoming marches, senators would, through placards, be called out by name to report to the full Senate on evidence received on industry underreporting. A sufficient number of such placards would get reported by the media.

The group Sunrise organized a Climate March on Washington for June 28, 2021. More such marches will be needed, marches incorporating skills from Women's March of 2017 organizers.

Each of these outfits deserves messaging toward future Climate Marches.

And, because the record indicates that turbo-capitalism has helped shoot us past the Paris Agreement warming standard, currently being planned is a great "economic restructuring" in America necessary to slow global warming. Thinkers behind this include Evan Weber, political director of the Sunrise group, who says,

> All that matters, in terms of continuing to have a livable planet, is whether we do what is necessary – which, according to science, is a massive, World War II-style mobilization to fully restructure our economy within our lifetimes.[2]

Vietnam Peace March, Washington DC, November 16, 1969

2 *The New Yorker*, "The Left Turn," May 31, 2021

Approximately 275,000 people marched here. Another 1.75 million marched in other U.S. cities

Women's March on Washington, January 21, 2017The Washington March drew approximately 475,000 people. As many as 4.5 million marched in other U.S. cities.

It has been noted that the Women's March in Washington that day in 2017 succeeded in mobilizing close to 2 percent of the entire U.S. population – women, children, and men. Nationwide, the Women's March mobilized 5 million marchers – 1.5 percent of the entire U.S. population. So, the Women's March succeeded in mobilizing considerably more people than did the Vietnam Moratorium, although the sentiments of anti-war in 1969 and pro-women's equality in 2017 ran roughly equally deeply; nearly everyone held them.

Much the same is now true of anti-climate change sentiment. Social media clearly helped the Women's March mobilize larger numbers than did Vietnam Moratorium, and the same advantage is available today to mobilize yet larger numbers for Climate Marches on Washington and around the country.

That is, following the numbers, a distinct possibility is 600,000 climate marchers in Washington and 6 million marchers nationwide.

It's the right thing to do.

Index

Symbols

A

B

C

D